MINITAB® MANUAL

MARIO F. TRIOLA
Dutchess Community College

TRIOLA STATISTICS SERIES

TWELFTH EDITION

Mario F. Triola
Dutchess Community College

PEARSON

Boston Columbus Indianapolis New York San Francisco Upper Saddle River
Amsterdam Cape Town Dubai London Madrid Milan Munich Paris Montreal Toronto
Delhi Mexico City Sao Paulo Sydney Hong Kong Seoul Singapore Taipei Tokyo

The author and publisher of this book have used their best efforts in preparing this book. These efforts include the development, research, and testing of the theories and programs to determine their effectiveness. The author and publisher make no warranty of any kind, expressed or implied, with regard to these programs or the documentation contained in this book. The author and publisher shall not be liable in any event for incidental or consequential damages in connection with, or arising out of, the furnishing, performance, or use of these programs.

Reproduced by Pearson from electronic files supplied by the author.

Publishing as Pearson, 75 Arlington Street, Boston, MA 02116.

ISBN-13: 978-0-321-83379-2
ISBN-10: 0-321-83379-1

2 3 4 5 6 EBM 16 15 14 13

www.pearsonhighered.com

PEARSON

Preface

This *Minitab Manual,* 12th edition, is a supplement to textbooks in the Triola Statistics Series (listed below), *Biostatistics for the Biological and Health Sciences* by Marc Triola, M.D. and Mario F. Triola, and *Statistical Reasoning for Everyday Life* by Bennett, Briggs, and Triola.

Triola Statistics Series

Elementary Statistics, Twelfth Edition

Essentials of Statistics, Fifth Edition

Elementary Statistics Using Excel, Fifth Edition

Elementary Statistics Using the TI-83/84 Calculator, Fourth Edition

This manual is based on Minitab Release 16, but it can also be used with other versions of Minitab, including Minitab Release 15 and the *Student Edition of Minitab.*

The associated textbooks are packaged with a CD-ROM that includes Minitab worksheets. The data sets in Appendix B of the textbook are included as Minitab worksheets on that CD-ROM. Those Minitab worksheets are also available on the web site http://www.aw.com/triola. Page 223 of this manual/workbook lists the names of the Minitab worksheets on the CD-ROM.

Here are major objectives of this manual/workbook and the Minitab software:

- Describe how Minitab can be used for the methods of statistics presented in the textbook. Specific and detailed procedures for using Minitab are included, along with examples of Minitab screen displays.

- Involve students with a spreadsheet type of software application, which will help them work with other spreadsheet programs they will encounter in their professional work.

- Incorporate an important component of computer usage without using valuable class time required for concepts of statistics.

- Replace tedious calculations or manual construction of graphs with computer results.

- Apply alternative methods, such as simulations, that are made possible with computer usage.

- Include topics, such as analysis of variance and multiple regression, that require calculations so complex that they realistically cannot be done without computer software.

It should be emphasized that this manual/workbook is designed to be a *supplement* to textbooks in the Triola Statistics Series; it is not designed to be a self-contained statistics textbook. It is assumed throughout this manual/workbook that the theory, assumptions, and procedures of statistics are described in the textbook that is used.

Chapter 1 of this supplement describes some of the important basics for using Minitab. Chapters 2 through 14 in this manual/workbook correspond to Chapters 2 through 14 in *Elementary Statistics*, 12th edition. However, the individual chapter *sections* in this manual/workbook generally do *not* match the sections in the textbook. Each chapter includes a description of the Minitab procedures relevant to the corresponding chapter in the textbook. This cross-referencing makes it very easy to use this supplement with the textbook.

Chapters include illustrations of Minitab procedures as well as detailed steps describing the use of those procedures. It would be helpful to follow the steps shown in these sections so the basic procedures will become familiar. You can compare your own computer display to the display given in this supplement and then verify that your procedure works correctly. You can then proceed to conduct the experiments that follow.

We welcome any comments or suggestions that you might have for improving this Minitab manual/workbook. Please send them to the Pearson Addison–Wesley Statistics Editor.

The author and publisher are very grateful to Minitab, Inc. for the continuing support and cooperation.

M.F.T.
Madison, CT
January, 2013

Contents

1

Basics of Minitab

1-1 Starting Minitab

Minitab Release 16 is the latest version available as of this writing. It can be purchased and downloaded online. As this manual was being written, students could purchase and download Minitab for $99.99, and they could rent Minitab for as little as $29.99 for a semester. (Go to www.Minitab.com, select Minitab 16, and select the Academic Pricing option.) In addition to Minitab Release 16 and earlier releases, the *Student Edition of Minitab R 14* can also be used. Although this manual/workbook is based on Minitab Release 16, it can be used for earlier releases of Minitab as well as the *Student Edition of Minitab*.

System Requirements Here are the system requirements: The processor must be PC with 1 GHz 32 or 64 bit, the operating system must be Windows XP, Vista, or Windows 7, you must have 512 MB of memory available in RAM, you must have 160 MB of disk space, and your display must have resolution of 1024 x 768 or higher

Starting Minitab: With Minitab installed, you can start it by clicking on the Minitab icon on the startup screen.

You can also start Minitab as follows:

1. Click on the taskbar item of **Start**.
2. Select **Programs**.
3. Select **Minitab 16** (or whatever Minitab release you are using).
4. Select **Minitab**.

Session/Data Windows After opening Minitab, you should see a display of its two main windows: the Session window and the Data window (as shown on the top of the following page). The Session window displays results and it also allows you to enter commands. The top of the display on the following page consists of the Session window. The Data window, shown on the bottom of the display on the following page, consists of a spreadsheet for entering data in columns. These spreadsheets are called *worksheets* in Minitab. Minitab allows you to have multiple worksheets in different Data windows.

Session and Data Windows

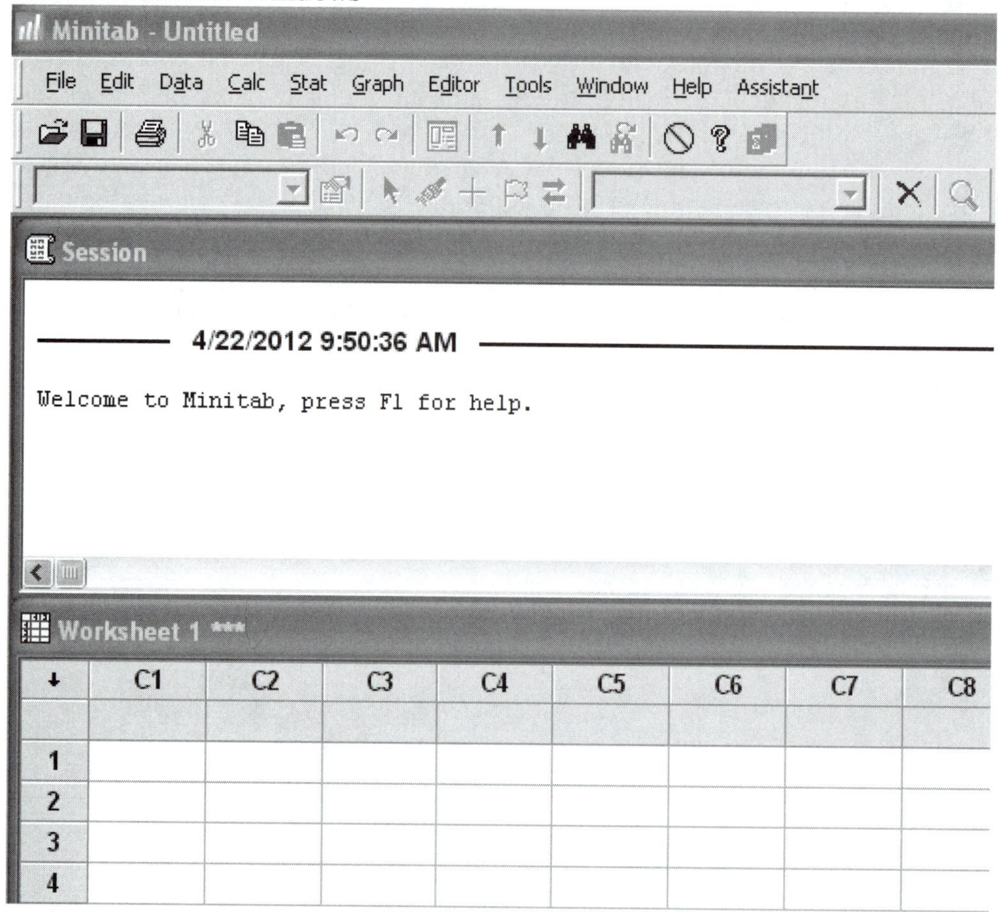

1-2 Downloading Worksheets

Data entered in Minitab can be saved as a worksheet (with an extension of .MTW) or a project file (with an extension of .MPJ).

Worksheet: Spreadsheet containing columns of data
Project: File containing all work created during a session, including worksheets, graphs, and session window output.

This manual/workbook and the stored Minitab data files all refer to *worksheets*.

Downloading Minitab Worksheets from the CD-ROM The CD-ROM included with the textbook includes Minitab worksheets containing the data sets in Appendix B from the textbook. For example, Data Set 13 in Appendix B of the textbook includes measurements from car crash tests, and that data set is available as the Minitab *worksheet* CRASH.MTW. The extension MTW shows us that this file is a Minitab worksheet, not a Minitab project. Here is the procedure for downloading a Minitab worksheet from the CD-ROM:

Downloading a Minitab Worksheet from the CD-ROM

The CD-ROM packaged with the textbook contains worksheets for the data sets in Appendix B of the textbook. See Appendix B in the textbook for the worksheet names.

1. Click on Minitab's main menu item of **File**.
2. Click on the subdirectory item of **Open Worksheet**.
3. You should now see a window like the one shown below.
4. In the "Look in" box at the top, select the location of the stored worksheets. For example, if the CD-ROM is in drive D, do this:

    - In the "Look in" box, select drive **D** (or whatever drive contains the CD-ROM).
    - Double click on the folder **App B Data Sets**.
    - Double click on the folder **Minitab**.
    - Click on the name of the worksheet that you want to open.

5. Click on the **Open** button.

1-3 Entering Data

In Minitab, there are three types of data: (1) *numeric* data (numbers), (2) *text* data (characters), and (3) *date/time* data. We will work mostly with numeric data. Also, data can be in the form of a column, a stored constant, or a matrix. We will work with columns of data. Shown below is a typical Minitab screen. We have entered data values in the first eight columns identified as C1, C2, C3, and so on. Because the entries in the first two columns consist of *text* data, the first column becomes C1-T and the second column becomes C2-T. We have also entered the column names of "CAR," "Size," and so on. (See Data Set 13 in Appendix B of the textbook.)

Entry of sample data is very easy. Click on the first cell on the desired column, type an entry, then press the **Enter** key. Then type a second entry and press the **Enter** key again. Continue until all of the sample data have been entered. If you make a mistake, simply click on the wrong entry and type the correct value and press **Enter**.

CRASH.MTW ***

↓	C1-T	C2-T	C3	C4	C5	C6	C7	C8
	CAR	SIZE	HIC	CHEST	FEML	FEMR	TTI	PLVS
1	Chev Aveo	Small	371	44	1188	1261	62	71
2	Honda Civic	Small	356	39	289	324	63	71
3	Mitsubishi Lancer	Small	275	37	329	446	35	45
4	VW Jetta	Small	544	54	707	1048	44	66
5	Hyundai Elantra	Small	326	39	602	1474	58	71
6	Kia Rio	Small	520	44	245	1046	64	84
7	Subaru Impreza	Small	443	42	334	455	50	53
8	Ford Fusion	Midsize	366	36	399	844	51	78
9	Nissan Altima	Midsize	287	53	317	713	53	53
10	Nissan Maxima	Midsize	255	43	301	133	44	59
11	Honda Accord	Midsize	249	42	297	236	59	55
12	Volvo S60	Midsize	502	52	810	687	49	67
13	VW Passat	Midsize	502	49	280	905	43	61
14	Toyota Camry	Midsize	505	41	411	547	42	57
15	Toyota Avalon	Large	342	32	215	752	36	55
16	Hyundai Azera	Large	698	45	1636	1202	48	75
17	Cadillac DTS	Large	346	41	738	772	61	75
18	Lincoln Town	Large	608	38	882	554	57	61
19	Dodge Charger	Large	216	37	937	669	63	53
20	Merc Gr Marq	Large	608	38	882	554	57	61
21	Buick Lucerne	Large	169	33	472	290	64	76

Naming columns: It is wise to enter a name for the different columns of data, so that it becomes easy to keep track of data sets with meaningful names. See the preceding display that includes the column names of "CAR," "SIZE," and so on.

1-4 Saving Data

After entering one or more columns of data, you can *save* all of the data as a Minitab worksheet, which is a file with an extension of .MTW. Use the following procedure.

Minitab Procedure for Saving a Worksheet

1. Click on the main menu item of **File**.
2. Click on **Save Current Worksheet As ….**
3. You will see a dialog box, such as the one shown below. In the "Save in" box, enter the location where you want the worksheet saved. In the "File name" box, enter the name of the file. If you omit the extension of .MTW, it will be automatically provided by Minitab.
4. Click on **Save**.

Note: For saved worksheets, Minitab is *forward* compatible, but not *backward* compatible. For example, a worksheet saved in Minitab Release 16 format is not compatible with Minitab Release 15 and earlier releases of Minitab, but a worksheet saved in a Minitab Release 14 format can be opened and used in Minitab 14, Minitab 15, Minitab 16, and later releases.

1-5 Printing

We have noted that Minitab involves a screen with two separate parts: The Session window at the top and the Data window at the bottom. The contents of the Session window and Data window can be printed as follows.

Printing the Session Window or Data Window

1. *Session window:* If you want to print the session window portion of the screen, click on the session window.

   Data window: If you want to print the data portion of the screen, click on any cell in the Data window.

2. Click on the main menu item of **File**.

3. Click on **Print Session Window** or **Print Worksheet**. (You will see only one of these two options. The option that you see is the result of the choice you made in Step 1. If you see "Print Worksheet" but you want to print the Session window, go back to Step 1 and click on the Session window portion of the screen.)

4. If you select Print Worksheet in Step 3, you will get a dialog box. Enter your preferences, then click **OK**.

Printing a Large Data Set in a Word Processor

Most of the data sets used with the Triola statistics series of textbooks are not so large that they cannot be printed on a few pages. If you do enter or somehow create a data set with thousands of values, printing would require many pages. One way to circumvent that problem is to move the data to a word processor where the values can be reconfigured for easier printing. Follow these steps.

1. With the data set displayed in the data window, click on the value at the top. Hold the mouse button down and drag the mouse to the bottom of the data set, then release it so that the entire list of values is highlighted.

2. Click on **Edit**, then click on **Copy Cells**.

3. Now go into your word processor and click on **Edit**, then **Paste**. The entire list of values will be in your word processing document where you can configure them as you please. For example, you might press **End**, then press the **space bar**, then press the **Del** key. The second value will be moved up to the top row. Repeat this process to rearrange the data in multiple rows and columns (instead of one really large column). For example, this approach was used to print on *one* page the 175 values in the column labeled CANS109 that is in the Minitab worksheet named CANS. The result is shown below. You can change the number of columns as you desire.

270 273 258 204 254 228 282 278 201 264 265 223 274 230 250 275 281 271 263 277
275 278 260 262 273 274 286 236 290 286 278 283 262 277 295 274 272 265 275 263
251 289 242 284 241 276 200 278 283 269 282 267 282 272 277 261 257 278 295 270
268 286 262 272 268 283 256 206 277 252 265 263 281 268 280 289 283 263 273 209
259 287 269 277 234 282 276 272 257 267 204 270 285 273 269 284 276 286 273 289
263 270 279 206 270 270 268 218 251 252 284 278 277 208 271 208 280 269 270 294
292 289 290 215 284 283 279 275 223 220 281 268 272 268 279 217 259 291 291 281
230 276 225 282 276 289 288 268 242 283 277 285 293 248 278 285 292 282 287 277
266 268 273 270 256 297 280 256 262 268 262 293 290 274 292

1-6 Command Editor and Transforming Data

Data may be *transformed* with operations such as adding a constant, multiplying by a constant, or using functions such as logarithm (common or natural), sine, exponential, or absolute value. For example, if you have a data set consisting of temperatures on the Fahrenheit scale (such as the body temperatures in the Minitab data set BODYTEMP included on the CD) and you want to transform the values to the Celsius scale, you can use the equation

$$C = \frac{5}{9}(F - 32)$$

Such transformations can be accomplished by using Minitab's *command editor* or by using Minitab's virtual calculator.

Using Command Editor to Transform Data

1. Click on the Session window portion of the screen. (The Session window is on the top portion; the Data window is the bottom portion.)
2. Click on **Editor**.
3. Click on **Enable Commands**.
4. You should now see **MTB >** and you can enter a command. The command

   LET C2 = (5/9)*(C1 – 32)

   tells Minitab to take the values in column C1, subtract 32 from each value, then multiply by the fraction 5/9. This expression corresponds to the above formula for converting Fahrenheit temperatures to the Celsius scale. The temperatures on the Fahrenheit scale of 98.6, 98.6, 98.0, 97.3, and 97.2 are converted to these temperatures on the Celsius scale: 37.0000, 37.0000, 36.6667, 36.2778, and 36.2222.

Using Minitab's Calculator to Transform Data

You can also click on **Calc**, then **Calculator** to get a virtual calculator that allows you to create new columns by performing operations with existing columns. The display below shows that column C3 will be the result of converting the temperatures in column C1 from the Fahrenheit scale to the Celsius scale. The expression shown below indicates that 32 is subtracted from each entry in column C1, then the result is multiplied by 5/9. Click **OK** when done.

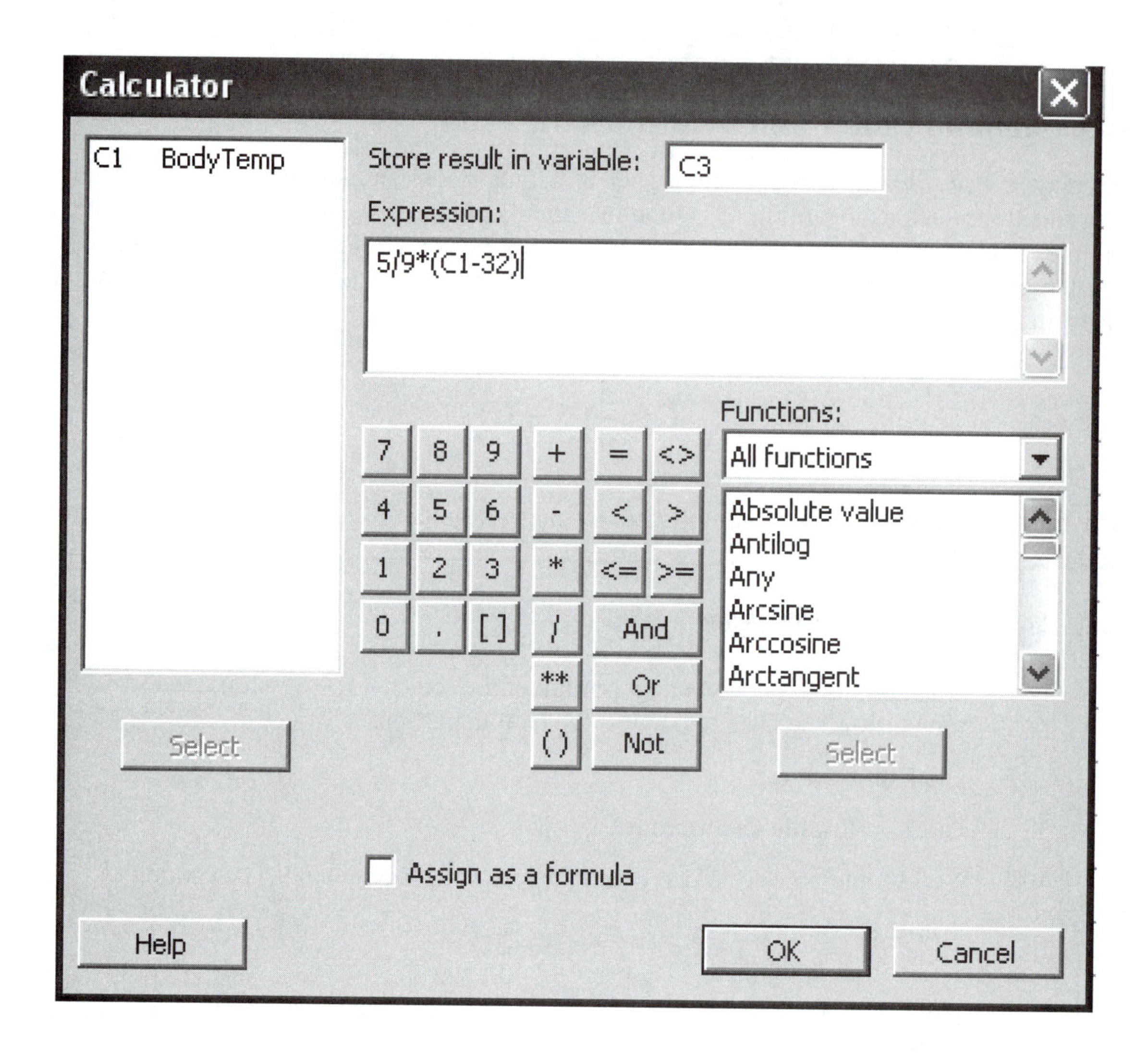

1-7 Closing Worksheets and Exiting Minitab

Closing worksheets: You can close the current *worksheet* by clicking on × located at the upper right corner of the *worksheet window*. If you click on ×, you will be told that the data in the worksheet will be removed and this action cannot be undone. You will be given the opportunity to save the data in a separate file, or you can click **No** to close the window. The Minitab program will continue to be active, even though you have just closed the current worksheet.

Exiting Minitab: Had enough for now? To exit or quit the Minitab program, click on × located in the extreme upper right corner. Another way to exit Minitab is to click on **File**, then click on **Exit**. In both cases, you will be asked if you want to save changes to the project before closing. This is your last chance to save items that you want saved. Assuming that you are finished and you want to exit Minitab, click **No**.

1-8 Exchanging Data with Other Applications

There may be times when you want to move data from Minitab to another application (such as Excel or STATDISK or Microsoft Word) or to move data from another application to Minitab. Instead of manually retyping all of the data values, you can usually transfer the data set directly. Given below are two ways to accomplish this. The first of the following two procedures is easier, if it works (and it often does work).

Method 1: Use Copy/Paste

1. With the data set displayed in the source program, use the mouse to highlight the values that you want to copy. (Click on the first value, then hold the mouse button down and drag it to highlight all of the values, then release the mouse button.)

2. While still in the source program, click on **Edit**, then click on **Copy**.

3. Now go into the software program that is your destination and click on **Edit**, then **Paste**. The entire list of values should reappear.

Method 2: Use Text Files

1. In the software program containing the original set of data, create a text file of the data. (In Minitab you can create a text file by clicking on **File**, then **Other Files**, then **Export Special Text**.)

2. Minitab and most other major applications allow you to import the text file that was created. (To import a text file into Minitab, click on **File**, then **Other Files**, then **Import Special Text**.)

CHAPTER 1 EXPERIMENTS: Basics of Minitab

1-1. ***Entering Sample Data*** When first experimenting with procedures for using Minitab, it's a good strategy to use a small data set instead of one that is large. If a small data set is lost, you can easily enter it a second time. In this experiment, we will enter a small data set, save it, retrieve it, and print it. The Body Temperatures data set from Data Set 2 in Appendix B of the Triola textbook includes these body temperatures, along with others:

98.6 98.6 98.0 98.0 99.0

a. Start Minitab and enter the above sample values in column C1. (See the procedure described in Section 1-4 of this manual/workbook.)

b. Save the worksheet using the worksheet name of TEMP (because the values are body temperatures). See the procedure described in Section 1-3 of this manual/workbook.

c. Print the data set by printing the display of the data window.

d. Exit Minitab, then restart it and retrieve the worksheet named TEMP. Save another copy of the same data set using the file name of TEMP2. Print TEMP2 and include the title at the top.

1-2. ***Opening a Worksheet*** The Minitab worksheet OSCR.MTW is Data Set 11 in Appendix B of the textbook. It lists ages of Oscar-winning actresses and actors at the time of the awards ceremonies. The worksheet is already stored on the CD-ROM that is included with the textbook. Open the worksheet OSCR.MTW (see Section 1-2 of this manual/workbook), then print its contents.

1-3. ***Generating Random Data*** In addition to entering or downloading data, Minitab can also *generate* data sets. In this experiment, we will use Minitab to simulate 50 births.

a. Select **Calc** from the main menu bar, then select **Random Data.**
b. Select **Integer**, because we want whole numbers.
c. Enter 50 for the number of values, enter the column (such as C1) to be used for the data, enter 0 for minimum, and enter 1 for the maximum, then click **OK**.
d. Examine the displayed values and count the number of times that 1 occurs. Record the result here: __________ If we stipulate that a 0 simulates the birth of a boy and 1 simulates the birth of a girl, this experiment simulates 50 births and results in the number of girls.

1-4. ***Generating Random Data*** Experiment 1-3 results in the random generation of the digits 0 and 1. Minitab can also generate sample data from other types of populations. In this experiment, we will use Minitab to simulate the random selection of 50 IQ scores.

(*continued*)

a. Select **Calc** from the main menu bar, then select **Random Data.**
b. Select **Normal**. (The normal distribution is described in the textbook.)
c. Enter 50 for the number of values, enter the column (such as C1) to be used for the data, enter 100 for the mean, and enter 15 for the standard deviation. (These statistics are described in the textbook.) Click **OK** and print the results.

1-5. ***Saving a Worksheet*** Listed below are pulse rates of females (from Data Set 1 in Appendix B in the textbook.) Manually enter the pulse rates in column C1 and name that column Pulse Rate. Save the worksheet with the name Pulse. Print the worksheet.

Female Pulse Rates
78 80 68 56 76 78 78 90 96 60 98 66 100 76 64 82 62 72 78 74
90 90 68 72 82 72 78 104 62 72 72 88 74 72 82 78 78 98 72 64

1-6. ***Saving a Worksheet*** Listed below are the data from the Chapter Problem for Chapter 3 in the textbook. Enter the data in columns of a Minitab worksheet, and give the columns meaningful names. Save the worksheet with the name CHIPS. Print the worksheet.

Table 3-1 Numbers of Chocolate Chips in Different Brands of Cookies

Chips Ahoy (regular)
22 22 26 24 23 27 25 20 24 26 25 25 19 24 20 22 24 25 25 20
23 30 26 20 25 28 19 26 26 23 25 23 23 23 22 26 27 23 28 24
Chips Ahoy (chewy)
21 20 16 17 16 17 20 22 14 20 19 17 20 21 21 18
20 20 21 19 22 20 20 19 16 19 16 15 24 23 14 24
Chips Ahoy (reduced fat)
13 24 18 16 21 20 14 20 18 12 24 23 28 18 18 19 22 21 22 16
13 20 20 23 24 20 17 20 19 21 27 16 24 19 23 25 14 18 15 19
Keebler
29 31 25 32 27 31 30 29 31 26 32 33 32 30 33 29 30
28 32 35 37 31 24 30 30 34 29 27 24 38 37 32 26 30
Hannaford
13 15 16 21 15 14 14 15 13 13 16 11
14 12 13 12 14 12 16 17 14 16 14 15

1-7. ***Transforming Data*** Experiment 1-1 results in saving the worksheet TEMP that contains 5 body temperatures in degrees Fahrenheit. Retrieve that data set, then proceed to transform the temperatures to the Celsius scale. Store the Celsius temperatures in column C2. (See Section 1-6 in this manual/workbook.) Print the worksheet containing the original Fahrenheit temperatures in column C1 and the corresponding Celsius temperatures in column C2.

2

Summarizing and Graphing Data

Important note: The topics of this chapter require that you use Minitab to enter data in a worksheet, retrieve worksheets, save worksheets, and print results, as described in Chapter 1 of this manual/ workbook.

2-1 Frequency Distributions

Section 2-2 in the textbook describes the construction of a table representing the *frequency distribution* for a set of data. Shown below are the full IQ scores of subjects from the low lead group, followed by Table 2-2 from the textbook. Table 2-2 is a frequency distribution summarizing the listed IQ scores. (The listed IQ scores are part of Data Set 5 in Appendix B, and they are included in the Minitab worksheet named IQLEAD.MTW.)

Full IQ Scores of Low Lead Group

70 85 86 76 84 96 94 56 115 97 77 128 99 80 118 86 141 88 96 96
107 86 80 107 101 91 125 96 99 99 115 106 105 96 50 99 85 88 120 93
87 98 78 100 105 87 94 89 80 111 104 85 94 75 73 76 107 88 89 96
72 97 76 107 104 85 76 95 86 89 76 96 101 108 102 77 74 92

Table 2-2 IQ Scores of Low Lead Group

IQ Score	Frequency
50 – 69	2
70 – 89	33
90 – 109	35
110 – 129	7
130 – 149	1

Minitab does not have a specific feature for generating a frequency distribution as in Table 2-2, but the following procedure can be very helpful in constructing such a table.

1. Enter the data in column C1 of the Minitab worksheet (or open a Minitab worksheet that includes the data).
2. Click on **Statistics,** then click on **Tables**.
3. Select **Tally Individual Variables**.
4. Enter **C1** (or the column name) in the box labeled “Variables.”
5. In the pop-up window, select the option of **Counts**.
6. Click **OK**.

If the preceding IQ scores are used, the first few lines of the result are as shown below.

IQFULL	Count
50	1
56	1
70	1
72	1
73	1
74	1
75	1
76	5
77	2
78	1
80	3
84	1
etc.	

With these tally counts, it is much easier to construct the frequency distribution in the format of the preceding Table 2-2. For example, it is easy to see from this tally list that there are 2 data values falling in the first class of 50–69.

Another way to obtain the frequency distribution is to generate a stemplot as described in Section 2-4 of this manual/workbook. The frequency counts needed for the frequency distribution can be identified by simply counting the number of digits to the right of each stem.

2-2 Histograms

Textbooks in the Triola textbook series describe histograms and provide detailed procedures for constructing them. It is noted that a histogram is an excellent device for exploring the *distribution* of a data set. Here is the Minitab procedure for generating a histogram.

1. Enter the data in a Minitab column, such as column C1. If the data are already stored in a Minitab worksheet, open the worksheet using the procedure described in Section 1-2 of this manual/workbook.

2. Click on the main menu item of **Graph**.

3. Select **Histogram** from the subdirectory.

4. Click on the option of **Simple**, then click **OK**.

5. In the dialog box, click on the column with the data so that the column label (such as C1) or the data name (IQFULL) appears in the box as shown.

Generating a Histogram

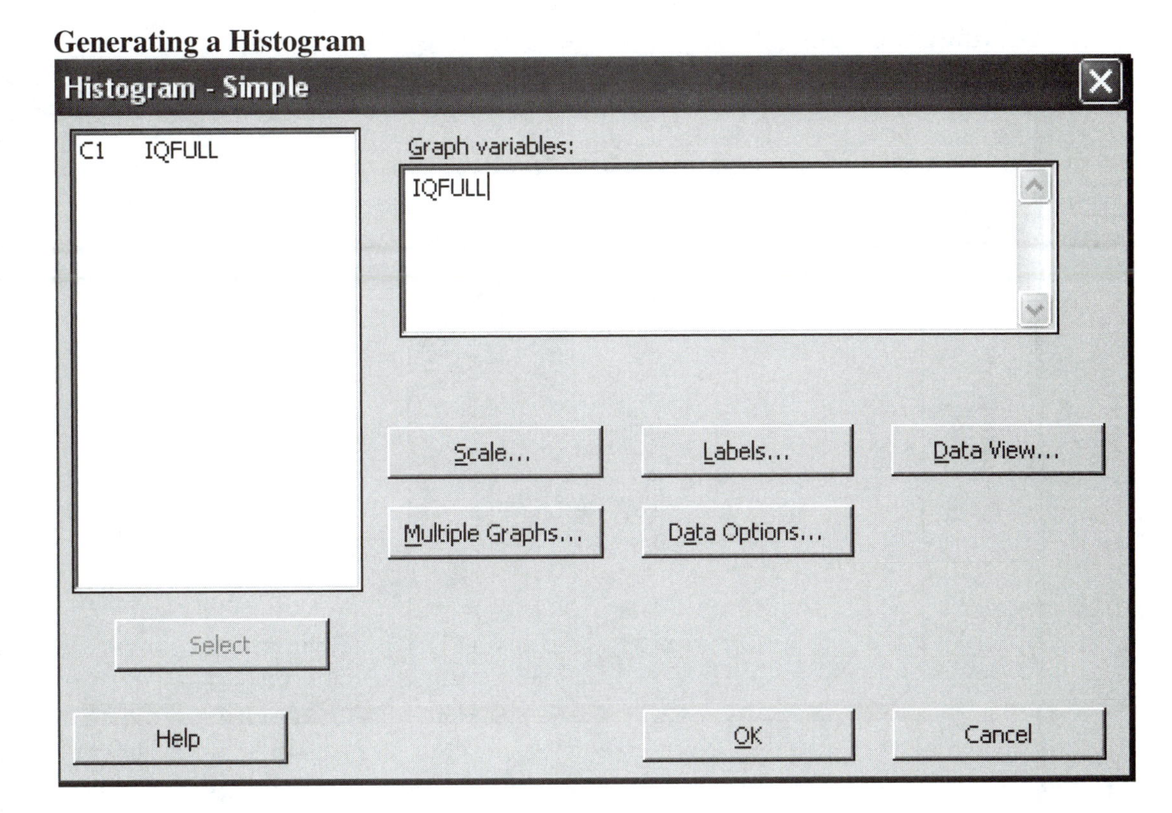

6. Using Minitab with the preceding IQ scores results in the following default histogram. Figure 2-2 in the textbook shows a different histogram with different class boundaries. The following defaullt Minitab-generated histogram can be edited by double-clicking on the attribute that you want to change. To change the horizontal scale to correspond to Figure 2-2 in the textbook, double-click on the *x*-axis and use the pop-up window to change the scale. For example, in the Edit Scale pop-up window, click on the **Binning** tab, select **Midpoint**, select **Midpoint/Cutpoint positions**, and enter the class midpoint values of 59.5, 79.5, 99.5, 119.5, 139.5. Click **OK** and the second graph on the following page will be displayed. This second graph corresponds closely to Figure 2-2 in the textbook. (Figure 2-2 was also edited to show red shading for the bars, and the values shown along the horizontal scale were edited to show class boundaries instead of class midpoints.)

Default Histogram

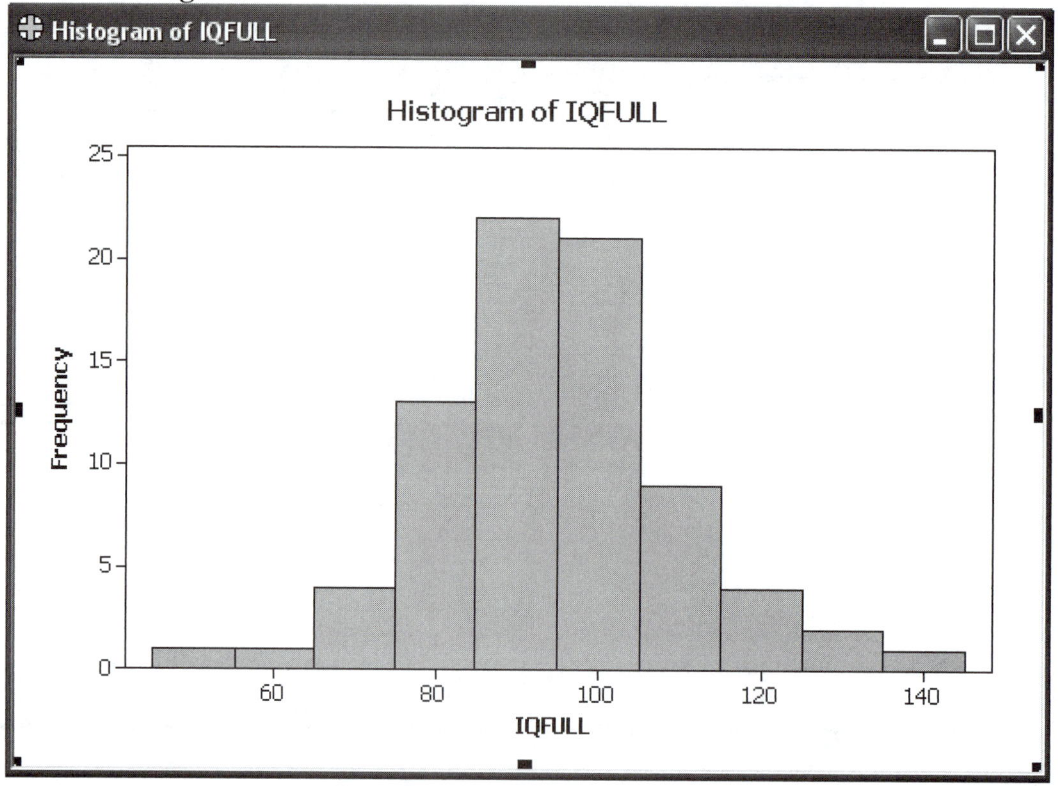

Histogram with Class Midpoints of 59.5, 79.5, 99.5, 119.5, 139.5

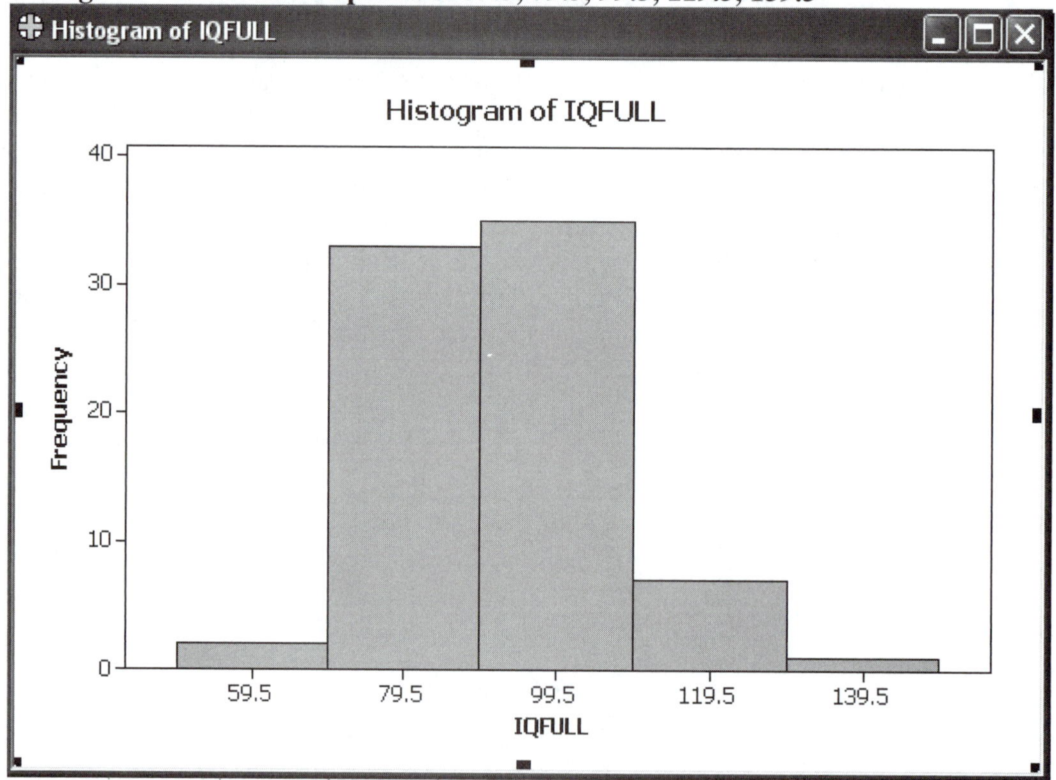

2-3 Normal Quantile Plots

The textbook points out that *normal quantile plots* are helpful in determining whether sample data appear to be from a population having a normal distribution. Section 6-6 in the textbook discusses methods for *assessing normality*, and Section 2-4 in the textbook includes a brief discussion of normal quantile plots. Minitab generates Probability Plots that can be interpreted using the same criteria for normal quantile plots.

Procedure for Generating a Probability Plot

1. Enter the data in a Minitab column, such as column C1. If the data are already stored in a Minitab worksheet, open the worksheet using the procedure described in Section 1-2 of this manual/workbook.

2. Click on the main menu item of **Graph**.

3. Select **Probability Plot** from the subdirectory.

4. Click on the option of **Single**, then click **OK**.

5. In the dialog box, identify the column to be used, then click **OK**.

Here is the probability plot obtained by using the full IQ scores listed in Section 2-1 of this manual/workbook.

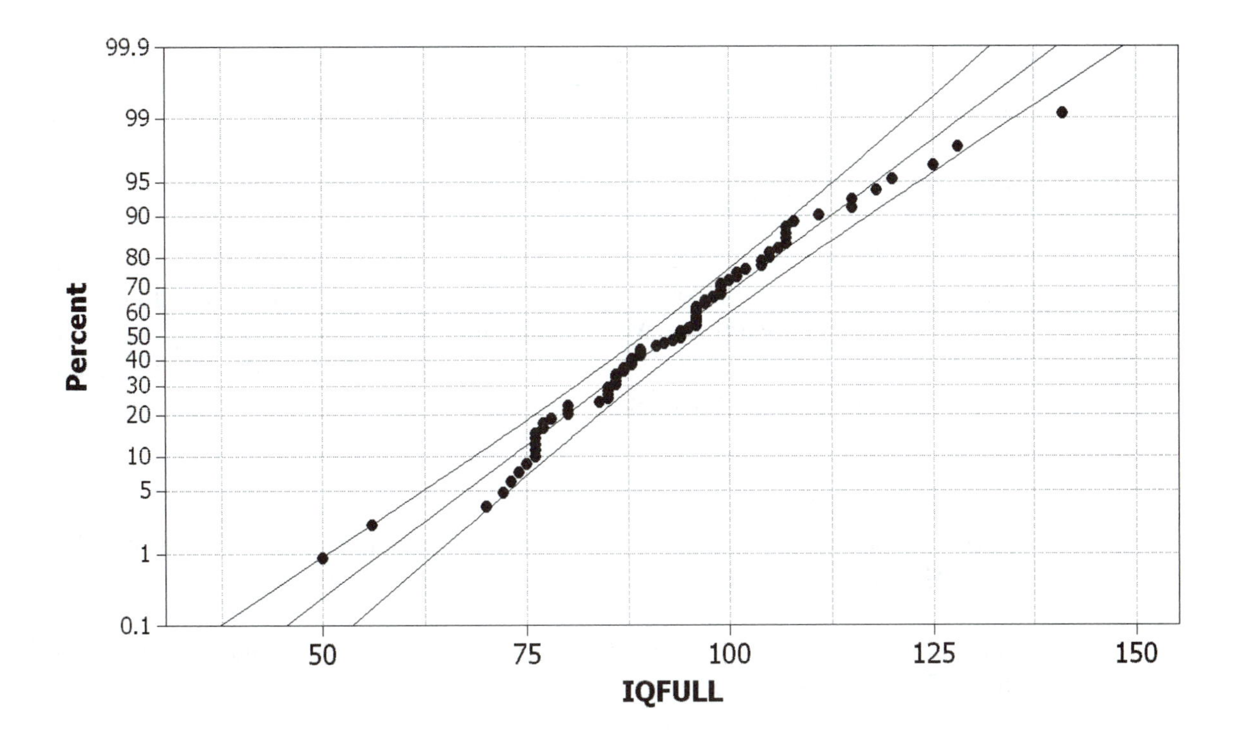

2-4 Scatterplots

The textbook describes a scatterplot (or scatter diagram). A scatterplot can be very helpful in seeing a relationship between two variables. The scatterplot shown below results from paired data consisting of the number of chirps per minute by a cricket and the temperature. This scatterplot shows that as the number of chirps increases, the corresponding temperature tends to be higher.

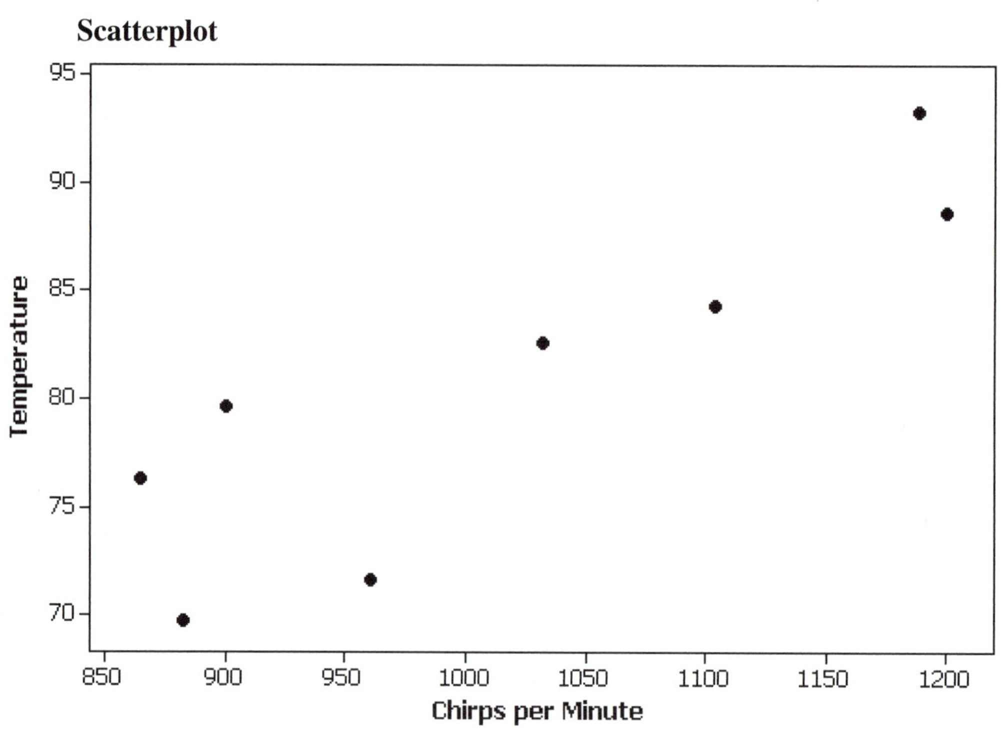

Here is the Minitab procedure for generating a scatterplot.

1. Given a collection of paired data, enter the values for one of the variables in column C1, and enter the corresponding values for the second variable in column C2. (You can also use any two Minitab columns other than C1 and C2.) Be careful to enter the two columns of data so that they are matched in the same way that they are paired in the original listing.

2. Click on **Graph** from the main menu.

3. Select the subdirectory item of **Scatterplot.** Select the option of **Simple**, then click **OK**.

4. In the dialog box, enter C1 in the box for the Y variable, then enter C2 in the box for the X variable (or vice versa).

5. Click **OK**.

Another procedure for generating a scatterplot is to select **Stat**, then **Regression**, then **Fitted Line Plot**. We will use this option later when we discuss the topics of correlation and regression. For now, we are simply generating the scatter diagram to visually explore whether there is an obvious pattern that might reveal some relationship between the two variables.

2-5 Time–Series Graph

Time–series data are data that have been collected at different points in time, and Minitab can graph such data so that patterns become easier to recognize. Here is the Minitab procedure for generating a time–series graph.

1. Given a collection of time–series data, enter the sequential values in column C1.
2. Select the main menu item of **Graph**.
3. Select the menu item of **Time Series Plot.**
4. Select the option of **Simple**.
5. A dialog box now appears. Enter C1 in the first cell under Y in the "Graph" list.
6. Enter a start time to be used for the horizontal axis. Click the **Options** button to enter a start time; click **Time/Scale** and proceed to select the desired time scale (such as "Calendar") along with a start value (such as 1980) and an increment value (such as 1). Click **OK** twice.

Listed below are the high values of the Dow Jones Industrial Average (DJIA) for each year from good job of showing the behavior of that stock market index.

1000 1024 1071 1287 1287 1553 1956 2722 2184 2791 3000 3169
3413 3794 3978 5216 6561 8259 9374 11568 11401 11350 10635 10454
10855 10941 12464 14198 13279 10580 11625

Time-Series Graph of DJIA High Values

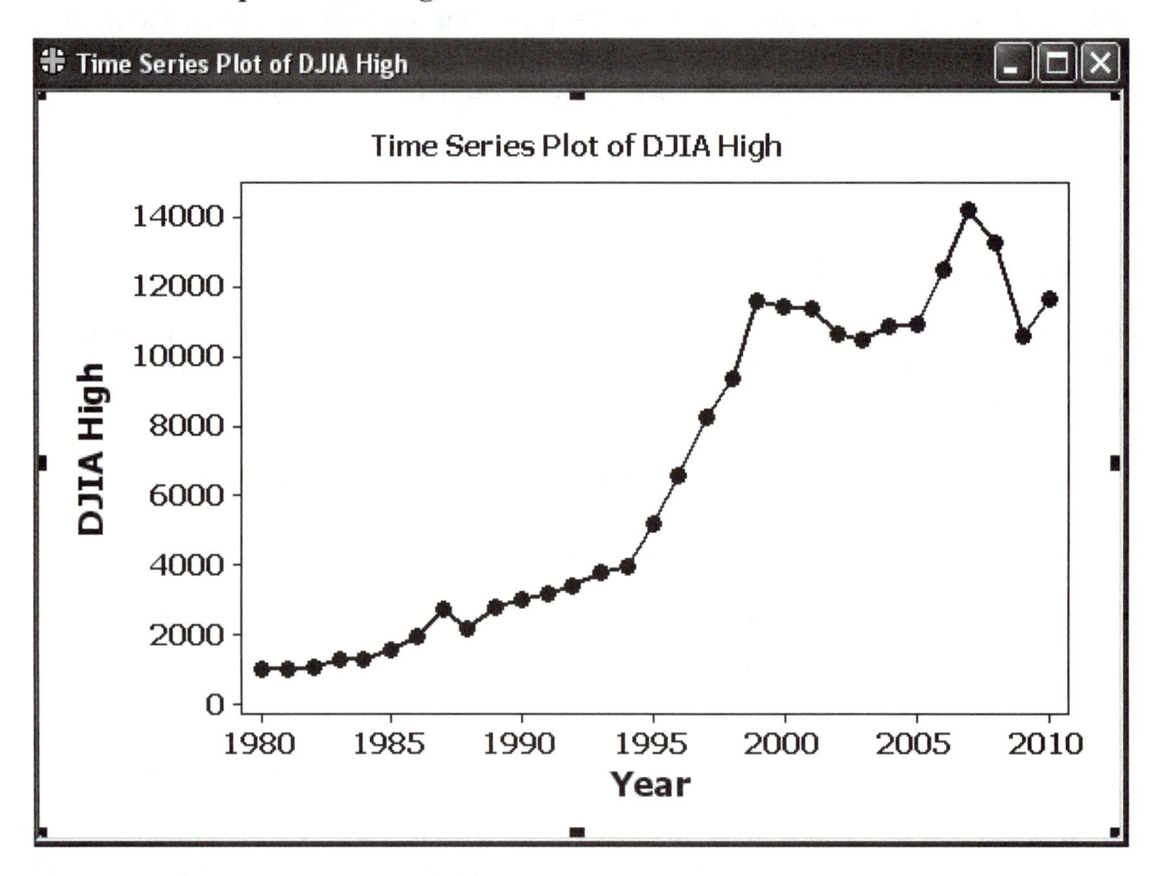

2-6 Dotplots

Here is the Minitab procedure for creating a dotplot.

1. Enter or retrieve the data into a Minitab column.

2. Click on the main menu item of **Graph**.

3. Select **Dotplot**.

4. Select the **Simple** dotplot and click **OK**.

5. The dialog box will list the columns of data at the left. Click on the desired column. You can edit a dotplot or any other graph by double-clicking on the attribute you want to modify. To change the scale, double-click on the axis and use the pop-up window.

6. Click **OK**.

Shown below is the default Minitab dotplot display of the full IQ scores listed in Section 2-1 of this manual/workbook. The second dotplot was modified to show the more convenient values of 50, 60, 70, . . . , 150 along the horizontal scale. This change in the numbers used for the horizontal scale was made by double-clicking on a number on the horizontal scale, selecting the **scale** tab, then manually entering the values of 50, 60, 70, . . . , 150 in the box labeled **Position of ticks**.

Default Dotplot

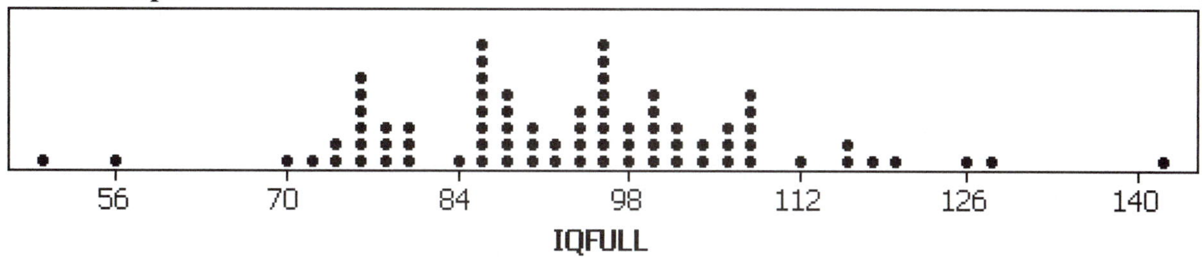

Dotplot With the Horizontal Scale Changed

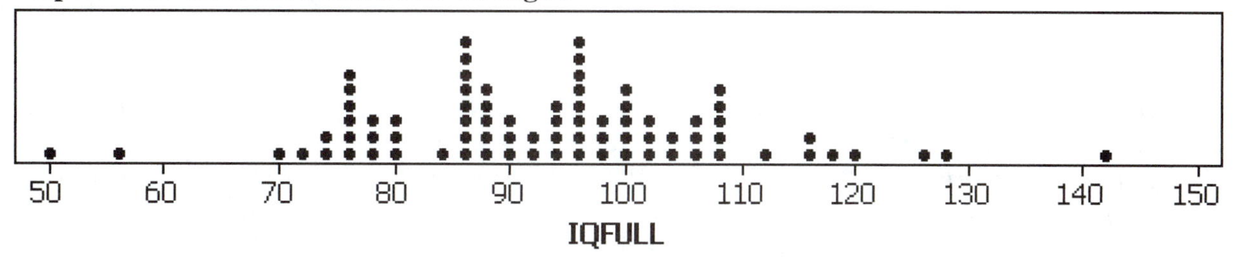

2-7 Stemplots

Minitab can provide stemplots (or stem-and-leaf plots). The textbook notes that a **stemplot** represents data by separating each value into two parts: the stem (such as the leftmost digit) and the leaf (such as the rightmost digit). Here is the Minitab procedure for generating a stemplot.

1. Enter or retrieve the data into a Minitab column.
2. Click on the main menu item of **Graph**.
3. Select **Stem-and-Leaf**.
4. The dialog box lists columns of data at the left. Click on the desired column.
5. Remove the check mark for "Trim outliers," so that outliers will be included.
6. Enter a desired value for the class interval, such as 10.
7. Click **OK**.

Shown below is the Minitab stemplot based on the IQ scores listed in Section 2-1 of this manual/workbook. The actual stemplot is highlighted below with a bold font. This stemplot is based on the use of 10 as the increment value.

Stem-and-Leaf Display: IQFULL

```
Stem-and-leaf of IQFULL  N  = 78
Leaf Unit = 1.0

  2     5    06
  2     6
 15     7    0234566666778
 35     8    00045555666677888999
(21)    9    123444566666667789999
 22    10    01124455677778
  8    11    1558
  4    12    058
  1    13
  1    14    1
```

Reading and Interpreting a Minitab Stemplot: In addition to the stemplot itself, Minitab also provides another column of data at the extreme left as shown above. In the above display, *the leftmost column represents cumulative totals*. The left column above shows that there are 2 sample values included between 50 and 59. The left column entry of (21) indicates that there are 21 data values in the row containing the median. The left-column entries *below* the median row represent cumulative totals from the bottom up, so that the 1 at the bottom indicates that there is one value between 140 and 149. This leftmost column of values is tricky to interpret, so ignoring it would not be the worst thing that could be done in your life.

2-8 Bar Graphs

Shown below is a typical Minitab-generated bar graph. The procedure for generating such a bar graph is as follows.

1. Enter the *names* of the categories (such as Democrats, Republicans, Independents) in column C1.
2. Enter the corresponding frequencies or percentage values in column C2.
3. Click on **Graph**.
4. Click on **Bar Chart.**
5. In the "Bars represent" box at the top, select the option of **Values from a table**.
6. Click on the **Simple** bar graph at the left, then click **OK**.
7. In the "Graph variables" box, enter C2 (or whatever column contains the frequencies or percentage values).

8. In the "Categorical variable" box, enter C1 (or whatever column contains the names of the categories).

9. Click **OK**.

Bar Graph

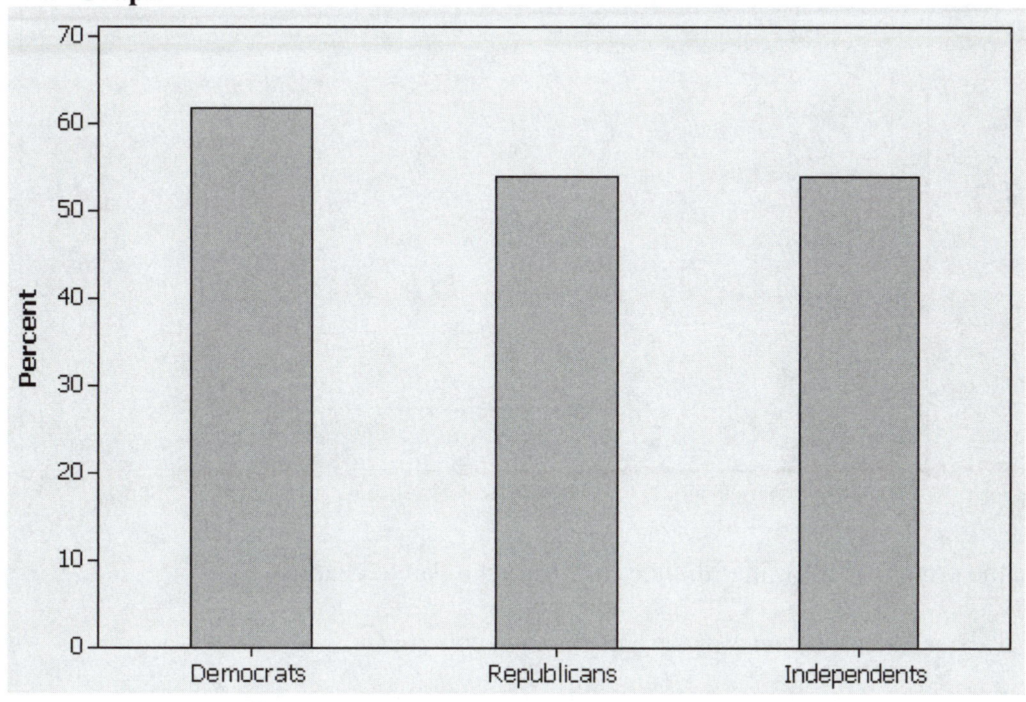

The textbook notes that a **multiple bar graph** has two or more sets of bars and is used to compare two or more data sets. Section 2-4 in the textbook includes an example of a multiple bar graph. The above procedure can be used for a multiple bar graph, but use the **Cluster** option in Step 6, not the option of Simple.

2-9 Pareto Charts

Section 2-4 of the textbook includes the Minitab-generated Pareto chart shown below. A Pareto chart is a bar graph for categorical data, with the bars arranged in order according to frequencies. The tallest bar is at the left, and the smaller bars are farther to the right, as shown below.

Pareto Chart

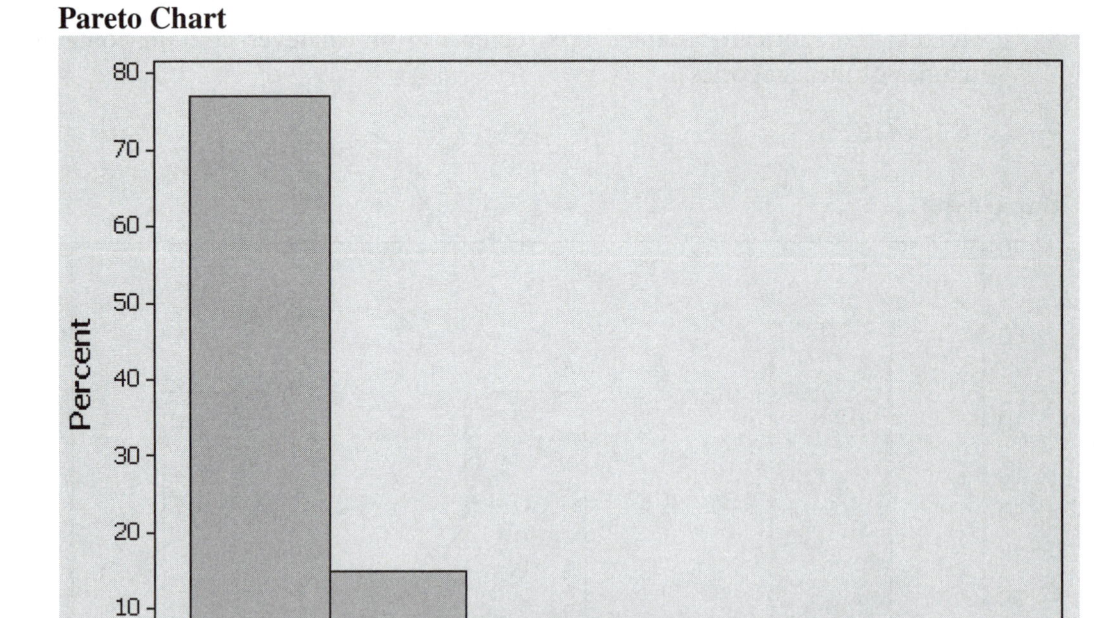

Here is the procedure for using Minitab to construct a Pareto chart:

1. Enter the *names* of the categories in column C1.

2. Enter the corresponding frequency counts in column C2.

3. Select the menu items of **Stat**, **Quality Tools**, and **Pareto Chart**.

4. When the dialog box is displayed, select the **Charts defect table** option and enter C1 in the labels box and C2 in the frequency box.

5. Click **OK**.

2-10 Pie Charts

To obtain a pie chart from Minitab, follow this procedure.

1. Enter the *names* of the categories in column C1.
2. Enter the corresponding frequency counts in column C2.
3. Select the main menu item of **Graph**.
4. Select **Pie Chart**.
5. When the dialog box is displayed, select **Chart values from a table** and enter C1 for the categorical variable box and enter C2 in the summary variables box.
6. Click **OK**. Shown below is the Minitab pie chart included in Section 2-4 of the textbook. (See Figure 2-13).

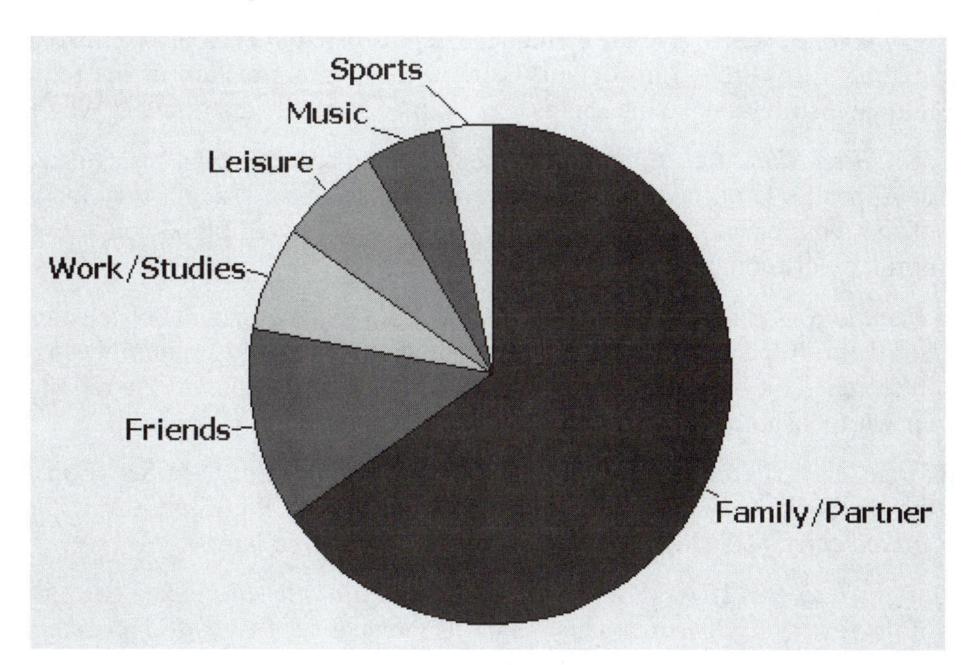

This chapter has described procedures for using Minitab to construct frequency distributions and a variety of different graphs. Boxplots are also helpful graphs, and they will be described in Chapter 3 of this manual/workbook.

CHAPTER 2 EXPERIMENTS: **Summarizing and Graphing Data**

Histograms. *In Experiments 2-1 through 2-8, use the data with Minitab to construct a histogram. Each experiment uses data from a Minitab worksheet corresponding to a data set in Appendix B of the textbook. It is not necessary to use a specific class width or class boundaries.*

2-1. *Pulse Rates of Males* Use the pulse rates of males listed in Data Set 1 in Appendix B of the textbook. Construct a histogram. Does the histogram appear to depict data having a normal distribution? Why or why not?

2-2. *Pulse Rates of Females* Use the pulse rates of females listed in Data Set 1 in Appendix B of the textbook. Construct a histogram. Does the histogram appear to depict data having a normal distribution? Why or why not?

2-3. *Earthquake Magnitudes* Use the earthquake magnitudes listed in Data Set 16 in Appendix B of the textbook. Construct a histogram. Using a loose interpretation of the requirements for a normal distribution, do the magnitudes appear to be normally distributed? Why or why not?

2-4. *Earthquake Depths* Use the earthquake depths listed in Data Set 16 in Appendix B of the textbook. Construct a histogram. Using a loose interpretation of the requirements for a normal distribution, do the depths appear to be normally distributed? Why or why not?

2-5. ***Male Red Blood Cell Counts*** Use the red blood cell counts of males listed in Data Set 1 in Appendix B of the textbook. Construct a histogram. Using a very loose interpretation of the requirements for a normal distribution, do the red blood cell counts appear to be normally distributed? Why or why not?

2-6. ***Female Red Blood Cell Counts*** Use the red blood cell counts of females listed in Data Set 1 in Appendix B of the textbook. Construct a histogram. Using a very loose interpretation of the requirements for a normal distribution, do the red blood cell counts appear to be normally distributed? Why or why not?

2-7. *Flight Arrival Times* Use the flight arrival times listed in Data Set 15 in Appendix B of the textbook. Construct a histogram. Which part of the histogram depicts flights that arrived early, and which part depicts flights that arrived late?

2-8. *Flight Taxi-Out Times* Use the flight taxi-out times listed in Data Set 15 in Appendix B of the textbook. Construct a histogram. If the quality of air traffic procedures is improved so that the taxi-out times vary much less, would the histogram be affected?

Scatterplots. *In Eperiments 2-9 through 2-12, use the given paired data from Appendix B to construct a scatterplot. Each experiment uses data from a Minitab worksheet corresponding to a data set in Appendix B of the textbook.*

2-9. *President's Heights* Refer to Data Set 12 in Appendix B and use the heights of U. S. Presidents and the heights of their main opponents in the election campaign. Does there appear to be a correlation?

2-10. ***Brain Volume and IQ*** Refer to Data Set 6 in Appendix B and use the brain volumes (cm^3) and IQ scores. A simple hypothesis is that people with larger brains are more intelligent and they have higher IQ scores. Does the scatterplot support that hypothesis?

2-11. ***Bear Chest Size and Weight*** Refer to Data Set 7 in Appendix B and use the measured chest sizes and weights of bears. Does there appear to be a correlation between those two variables?

2-12. ***Coke Volume and Weight*** Refer to Data Set 19 in Appendix B and use the volumes and weights of regular Coke. Does there appear to be a correlation between volume and weight? What else is notable about the arrangement of the points, and how can it be explained?

Time Series Graphs. *In Experiments 2-13 and 2-14, construct the time series graph.*

2-13. *Harry Potter* Listed below are the gross amounts (in millions of dollars) earned from box office receipts for the movie *Harry Potter and the Half-Blood Prince.* The movie opened on a Wednesday, and the amounts are listed in order for the first 14 days of the movie's release. What is an explanation for the fact that the three highest amounts are the first, third, and fourth values listed?

58 22 27 29 21 10 10 8 7 9 11 9 4 4

2-14. *Home Runs* Listed below are the numbers of home runs in major league baseball for each year beginning with 1990 (listed in order by row). Is there a trend?

3317 3383 3038 4030 3306 4081 4962 4640 5064 5528
5693 5458 5059 5207 5451 5017 5386 4957 4878 4655

Dotplots. *In Experiments 2-15 and 2-16, construct the dotplot.*

2-15. *Coke Volumes* Refer to Data Set 19 in Appendix B and use the volumes of regular Coke. Does the configuration of the points appear to suggest that the volumes are from a population with a normal distribution? Why or why not? Are there any outliers?

2-16. *Car Pollution* Refer to Data Set 14 in Appendix B and use the greenhouse gas emissions (GHG) from the sample of cars. Does the configuration of the points appear to suggest that the amounts are from a population with a normal distribution? Why or why not?

Stemplots. *In Experiments 2-17 and 2-18, refer to the data and use Minitab to construct a stemplot.*

2-17. *Car Crash Tests* Refer to Data Set 13 in Appendix B and use the 21 pelvis (PLVS) deceleration measurements from the car crash tests. Is there strong evidence suggesting that the data are *not* from a population having a normal distribution?

2-18. *Car Braking Distances* Refer to Data Set 14 in Appendix B and use the 21 braking distances (ft). Are there any outliers? Is there strong evidence suggesting that the data are *not* from a population having a normal distribution?

Pareto Charts. *In Experiments 2-19 and 2-20, construct the Pareto chart.*

2-19. *Awful Sounds* In a survey, 1004 adults were asked to identify the most frustrating sound that they hear in a day. 279 chose jackhammers, 388 chose car alarms, 128 chose barking dogs, and 209 chose crying babies (based on data from Kelton Research).

2-20. *School Day* Here are weekly instruction times for school children in different countries: 23.8 hours (Japan); 26.9 hours (China); 22.2 hours (U.S.), 24.6 hours (U.K.); 24.8 hours (France). What do these results suggest about education in the United States?

Pie Charts. *In Experiments 2-21 and 2-22, construct the pie chart.*

2-21. *Awful Sounds* Use the data from Experiment 2-19.

2-22. *School Day* Use the data from Experiment 2-20. Does it make sense to use a pie chart for the given data?

2–23. ***Working with Your Own Data***

Through observation or experimentation, collect your own set of sample values. Obtain at least 40 values and try to select data from an interesting population. Use Minitab to generate graphs suitable for describing the distribution of the data. Describe the data and any notable or important characteristics.

3

Statistics for Describing, Exploring, and Comparing Data

Important note: The topics of this chapter require that you use Minitab to enter data, download worksheets of data, save worksheets, and print results. These functions are described in Chapter 1 of this manual/ workbook. Be sure to understand those functions from Chapter 1 before beginning this chapter.

The textbook notes that these characteristics of a data set are extremely important: center, variation, distribution, outliers, and changing characteristics of data over time. Chapter 2 of this manual/workbook presented a variety of graphs that are helpful in learning about the distribution of a data set. This chapter presents methods for learning about center, variation, and outliers.

3-1 Descriptive Statistics

The Chapter Problem for Chapter 2 in the textbook includes the following full IQ scores of subjects from the low lead group, included in Data Set 5 in Appendix B, and included in the Minitab worksheet named IQLEAD.MTW.

Full IQ Scores of Low Lead Group

70 85 86 76 84 96 94 56 115 97 77 128 99 80 118 86 141 88 96 96
107 86 80 107 101 91 125 96 99 99 115 106 105 96 50 99 85 88 120 93
87 98 78 100 105 87 94 89 80 111 104 85 94 75 73 76 107 88 89 96
72 97 76 107 104 85 76 95 86 89 76 96 101 108 102 77 74 92

Given a data set, such as the above list of IQ scores, we can use Minitab to obtain descriptive statistics, including the mean, standard deviation, and quartiles. Here is the Minitab procedure.

1. Enter or open the data in a Minitab column, such as C1. (See Section 1-2.)
2. Click on the main menu item of **Stat**.
3. Click on the subdirectory item of **Basic Statistics**.
4. Select **Display Descriptive Statistics**.
5. A "Display Descriptive Statistics" dialog box will pop up. In the "Variables" box enter the column containing the data that you are investigating. You can manually enter the column, such as C1, or the name of the column, or you can click on the desired column displayed at the left.
    - You can also click on the **Graphs** bar to generate certain graphs, including a histogram and boxplot.
    - You can also click on the **Statistics** bar to select the particular statistics that you want included among the results.
6. Click **OK**.

If you use the IQ scores listed on the preceding page and follow the above procedure, you will get a Minitab display like the one shown below. For this display, the author clicked on the **Statistics** bar and selected the statistics that are particularly relevant to discussions in the textbook.

Descriptive Statistics: IQFULL

```
          Total
Variable  Count   Mean  StDev  Variance  Minimum     Q1  Median      Q3
IQFULL       78  92.88  15.34    235.45    50.00  84.75   94.00  101.25

Variable  Maximum  Range
IQFULL     141.00  91.00
```

From the above Minitab display, we obtain the following important descriptive statistics (expressed by applying the round-off rule of using one more decimal place than in the original data values).

- Number of data values: 78
- Mean: $\overline{x} = 92.9$
- Standard deviation: $s = 15.3$
- Variance: $s^2 = 235.5$
- Range: 91.0
- Five-number-summary:
  - Minimum: 50
  - First quartile Q_1: 84.8
  - Median: 94.0
  - Third quartile Q_3: 101.3
  - Maximum: 141

Quartiles: The textbook uses a simplified procedure for finding quartiles, so the quartiles Q_1 and Q_3 found with Minitab may differ slightly from those found by using the procedure described in the textbook. For the above IQ scores, Minitab's quartiles of 84.8 and 101.3 differ slightly from the values of 85.0 and 101.0 that would be obtained by using the method described in the textbook.

3-2 z Scores

The textbook describes z scores (or standard scores). For sample data with mean $\bar{x}$ and standard deviation s, the z score can be found for a sample value x by computing

$$z = \frac{x - \bar{x}}{s} \quad \text{or} \quad z = \frac{x - \mu}{\sigma}$$

Here is the Minitab procedure for finding z scores corresponding to sample values. (This procedure uses column C1, but any other Minitab column can be used instead.)

1. Enter a list of data values in column C1.
2. Click on the main menu item of **Calc**.
3. Select **Standardize** from the subdirectory.
4. A dialog box should appear. Enter C1 for the input column and C2 (or any other specific column) for the column in which the z-score results will be stored.
5. There are buttons for different calculations, but be sure to select the button for "Subtract mean and divide by standard deviation."
6. Click **OK**, and column C2 will magically appear with the z score equivalent of each of the original sample values. Because column C2 is now available as a set of data, you can explore it using Descriptive Statistics, Histogram, Dotplot, and so on.

Shown below are the first few rows from the Minitab display of the z scores for the above list of IQ scores. The column label of "z score" was manually entered.

IQFULL	z score
70	-1.49139
85	-0.51384
86	-0.44867
76	-1.10037
84	-0.57901

3-3 Boxplots

Section 3-4 in the textbook describes the construction of boxplots based on the minimum value, maximum value, median, and quartiles Q_1 and Q_3. *Note:* The values of Q_1 and Q_3 generated by Minitab may be slightly different from the values obtained by using the procedure described in the textbook. Here is the Minitab procedure for generating boxplots.

1. Enter the data or open a worksheet with data.
2. Click on the main menu item of **Graph**.
3. Select the subdirectory item of **Boxplot**.
4. Select **One Y Simple** for one boxplot or select **Multiple Y's Simple** for boxplots of two or more data sets.
5. In the dialog box, click on the column(s) containing the data for which the boxplot will be produced.
6. Click **OK**.

Shown below are the two boxplots for the pulse rates of females and males listed in Data Set 1 from Appendix B in the textbook. Boxplots are particularly useful for comparing data sets. The asterisks are used to identify data values that appear to be outliers. By comparing the two boxplots, we see that the pulse rates of females appear to be somewhat higher than those of males.

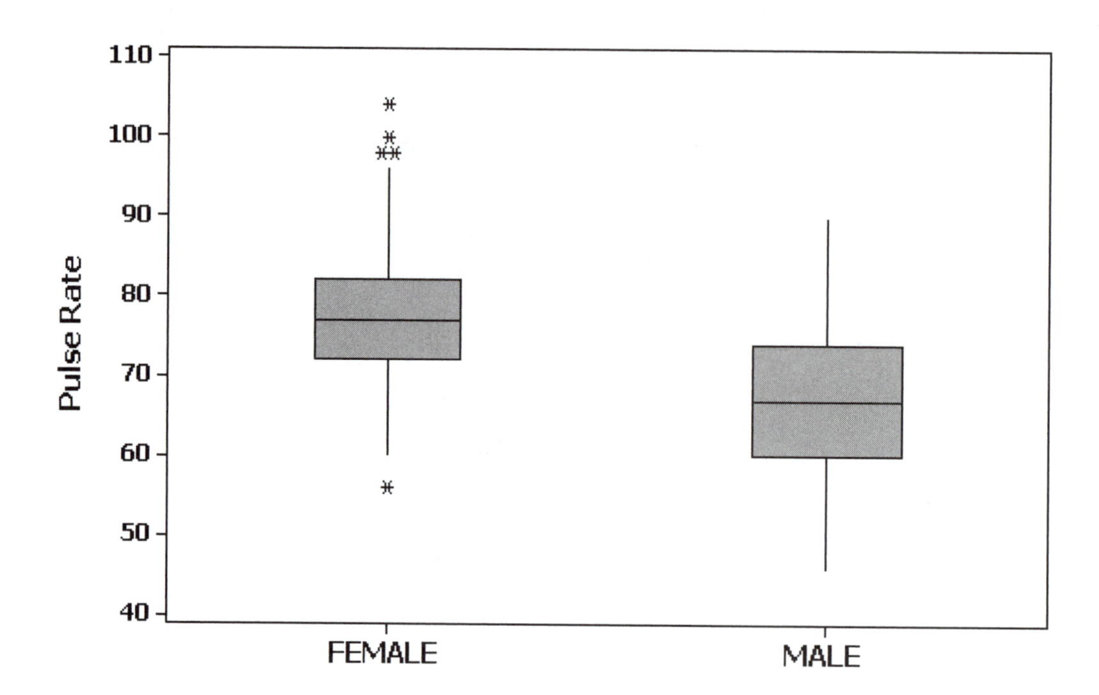

3-4 Sorting Data

There is often a need to *sort* data. Data are sorted when they are arranged in order from low to high (or from high to low). Identification of outliers becomes easier with a sorted list of values. Here is how we can use Minitab to sort data.

1. Enter the data or open a worksheet with data.

2. Click on the main menu item of **Data.**

3. Click on the subdirectory item of **Sort**.

4. You will now see a dialog box like the one shown below. In the "Sort column(s)" box, enter the column that you want to sort. In the "By column" box, enter the column to be used as the basis for sorting (this is usually the column containing the original unsorted data). In the "Column(s) of current worksheet" box, enter the column where you want the sorted data to be placed. (You can place the sorted data in the same column as the original data.) The entries shown below tell Minitab to sort the pulse rate data (in column C2) and store the sorted list in column C15. Click **OK** after completing the dialog box.

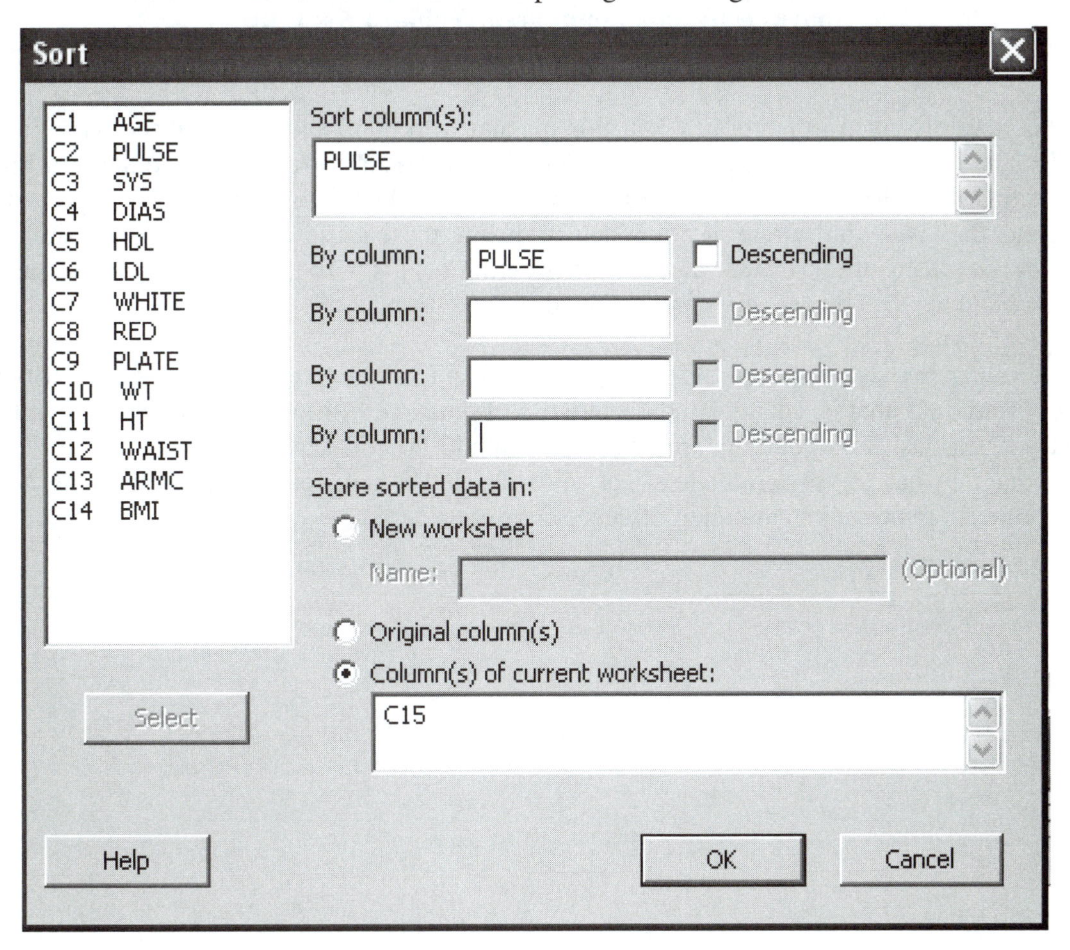

3-5 Outliers

Minitab does not have a specific function for identifying outliers, but some of the functions described in this chapter can be used. One easy procedure for identifying outliers is to *sort* the data as described in Section 3-4 of this manual/workbook, then simply examine the values near the beginning of the list or the end of the list. Consider a value to be an outlier if it is very far away from almost all of the other data values.

Another way to identify outliers is to construct a boxplot. We have noted that Minitab uses asterisks to depict outliers that are identified with a specific criterion that uses the *interquartile range* (IQR). The interquartile range IQR is as follows:

$$\text{IQR} = Q_3 - Q_1$$

Minitab identifies data values as outliers and denotes them with asterisks in a boxplot by applying these same criteria given in Part 2 of Section 3-4 from the textbook:

A data value is an outlier if it is ...

above Q_3 by an amount greater than $1.5 \times$ IQR
or **below Q_1 by an amount greater than $1.5 \times$ IQR.**

See the boxplots given in Section 3-3 of this manual/workbook and note that the pulse rates of females are depicted with a boxplot that includes four asterisks corresponding to the four highest pulse rates of 98, 98, 100, 104. There is also an asterisk at the bottom of the boxplot, and it represents the lowest pulse rate of 56. When analyzing these pulse rates, it is a good strategy to consider the effects of those five values by considering results with those values included and the results obtained when those values are excluded.

Remember that important characteristics of data sets are center, variation, distribution, outliers, and time (that is, changing characteristics of data over time). Outliers are important for a variety of reasons. Some outliers are errors that should be corrected. Some outliers are correct values that may have a very dramatic effect on results and conclusions, so it is important to know when outliers are present so that their effects can be considered.

CHAPTER 3 EXPERIMENTS:
Statistics for Describing, Exploring, and Comparing Data

3–1. ***Comparing Heights of Men and Women*** In this experiment we use two small data sets as a quick introduction to using some of the basic Minitab features. (When beginning work with new software, it is wise to first work with small data sets so that they can be entered quickly if they are lost or damaged.) The data listed below are measured heights (cm) of random samples of men and women (taken from Data Set 1 in Appendix B of the textbook).

Men	178.8	177.5	187.8	172.4	181.7	169.0
Women	163.7	165.5	163.1	166.3	163.6	170.9

a. Find the indicated characteristics of the heights of *men* and enter the results below.

Center: Mean: _____ Median: _____

Variation: St. Dev.:_____ Range: _____

5-Number Summary: Min.:_____ Q_1:_____ Q_2:_____ Q_3:_____ Max.:_____

Outliers: ____________________

b. Find the characteristics of the heights of *women* and enter the results below.

Center: Mean: _____ Median: _____

Variation: St. Dev.:_____ Range: _____

5-Number Summary: Min.:_____ Q_1:_____ Q_2:_____ Q_3:_____ Max.:_____

Outliers: ____________________

c. Compare the results from parts a and b.

____

____

____

3–2. ***Working with Larger Data Sets*** Repeat Experiment 3–1, but use the sample data for all 40 males and 40 females included in Data Set 1 that is found in Appendix B of the textbook. Instead of manually entering the 80 individual heights (which would be no fun at all), open the worksheets MBODY and FBODY that are found among the Minitab data sets on the CD–ROM that is included with the textbook.

a. Find the indicated characteristics of the heights of *men* and enter the results below.

Center: Mean: _____ Median: _____

Variation: St. Dev.:_____ Range: _____

5-Number Summary: Min.:_____ Q_1:_____ Q_2:_____ Q_3:_____ Max.:_____

Outliers: ___________________

b. Find the characteristics of the heights of *women* and enter the results below.

Center: Mean: _____ Median: _____

Variation: St. Dev.:_____ Range: _____

5-Number Summary: Min.:_____ Q_1:_____ Q_2:_____ Q_3:_____ Max.:_____

Outliers: ___________________

c. Compare the results from parts a and b.

____

____

____

d. Are there are notable differences observed from the complete sets of sample data that could not be seen with the smaller samples listed in Experiment 3–1? If so, what are they?

____

____

____

3–3. ***Boxplots*** Use the same sets of data used in Experiment 3–2 and print boxplots for the heights of the 40 men and the heights of the 40 women. Include both boxplots in the same window so that they can be compared. Do the boxplots suggest any notable differences in the two sets of sample data?

____

____

Interpret the *asterisk* that appears in the Minitab display for the boxplot depicting the heights of the men.

____

____

3–4. ***Effect of Outlier*** In this experiment we will study the effect of an *outlier*. Use the same heights of *men* used in Experiment 3-1, but change the first entry from 178.8 cm. to 1788 cm. (This type of mistake often occurs when the key for the decimal point is not pressed with enough force.) The outlier of 1788 cm is clearly a mistake, because a male with of height of 1788 cm would be 59 feet tall, or about six stories tall. Although this outlier is a mistake, outliers are sometimes correct values that differ substantially from the other sample values.

Men (cm): **1788** 177.5 187.8 172.4 181.7 169.0

Using this modified data set with the height of 178.8 cm changed to be the outlier of 1788 cm, find the following.

Center: Mean: _____ Median: _____

Variation: St. Dev.:_____ Range: _____

5-Number Summary: Min.:_____ Q_1:_____ Q_2:_____ Q_3:_____ Max.:_____

Outliers: ____________________

Based on a comparison of these results to those found in Experiment 3–1, how is the mean affected by the presence of an outlier?

____

How is the median affected by the presence of an outlier?

____

How is the standard deviation affected by the presence of an outlier?

____

3–5. ***Describing Data*** Use the 40 BMI (body mass index) indices of females from Data Set 1 in Appendix B from the textbook, and enter the results indicated below. (The Minitab worksheet is FBODY.MTW.)

Center: Mean: _____ Median: _____

Variation: St. Dev.:_____ Range: _____

5-Number Summary: Min.:_____ Q_1:_____ Q_2:_____ Q_3:_____ Max.:_____

Outliers: ____________________

3–6. ***Describing Data*** Use the 40 BMI (body mass index) indices of males from Data Set 1 in Appendix B from the textbook, and enter the results indicated below. (The Minitab worksheet is MBODY.MTW.)

Center: Mean: _____ Median: _____

Variation: St. Dev.:_____ Range: _____

5-Number Summary: Min.:_____ Q_1:_____ Q_2:_____ Q_3:_____ Max.:_____

Outliers: ____________________

Compare the BMI indices of females (from Experiment 3-5) and males.

____

____

3–7. ***Boxplots*** Use the BMI indices of females (see Experiment 3-5) and use the BMI indices of males (see Experiment 3-6). Print their two boxplots together in the same display. (*Hint:* Because the two lists are in different data sets, use Copy/Paste to configure the two lists of values so that they are in the same Sample Editor window.) Do the boxplots suggest any notable differences in the two sets of sample data?

____

____

3–8. ***Sorting Data*** Use the earthquake magnitudes listed in Data Set 16 from Appendix B in the textbook. (The Minitab worksheet is QUAKE.MTW.) *Sort* that data set by arranging the magnitudes from lowest to highest. List the first ten values in the sorted column.

____

3–9. ***z* Scores** Open the Minitab worksheet CANS.MTW, then use Minitab to find the z score corresponding to each of the 175 values for the cans that are 0.0109 in. thick. [Each value is an axial load, which is the weight (in pounds) that the can supports before being crushed.] Store the 175 z scores in column C3, then use column C3 to find the following.

Center: Mean: _____ Median: _____

Variation: St. Dev.:_____ Range: _____

5-Number Summary: Min.:_____ Q_1:_____ Q_2:_____ Q_3:_____ Max.:_____

Outliers: ____________________

What is notable about the above results? Specifically, what is notable about the value of the mean and standard deviation?

____

Will the same mean and standard deviation be obtained for *any* set of sample data? Explain how you arrived at your answer.

____

____

3–10. ***Combining Data*** Open the Minitab worksheet M&M, which consists of six different columns of data. Use **Data/Stack** to make a copy of the data all stacked together in column C7 of the current worksheet.

a. Find the following results for the combined data in column C7.

Center: Mean: _____ Median: _____

Variation: St. Dev.:_____ Range: _____

5-Number Summary: Min.:_____ Q_1:_____ Q_2:_____ Q_3:_____ Max.:_____

Outliers: ____________________

b. Print a boxplot of the data set.

c. Describe the important characteristics of the data set. Be sure to address the nature of the distribution, measures of center, measures of variation, and any other important and notable features.

____

____

3–11. ***Statistics from a Dotplot*** Shown below is a Minitab dotplot. Identify the values represented in this graph, enter them in column C1, then find the indicated results.

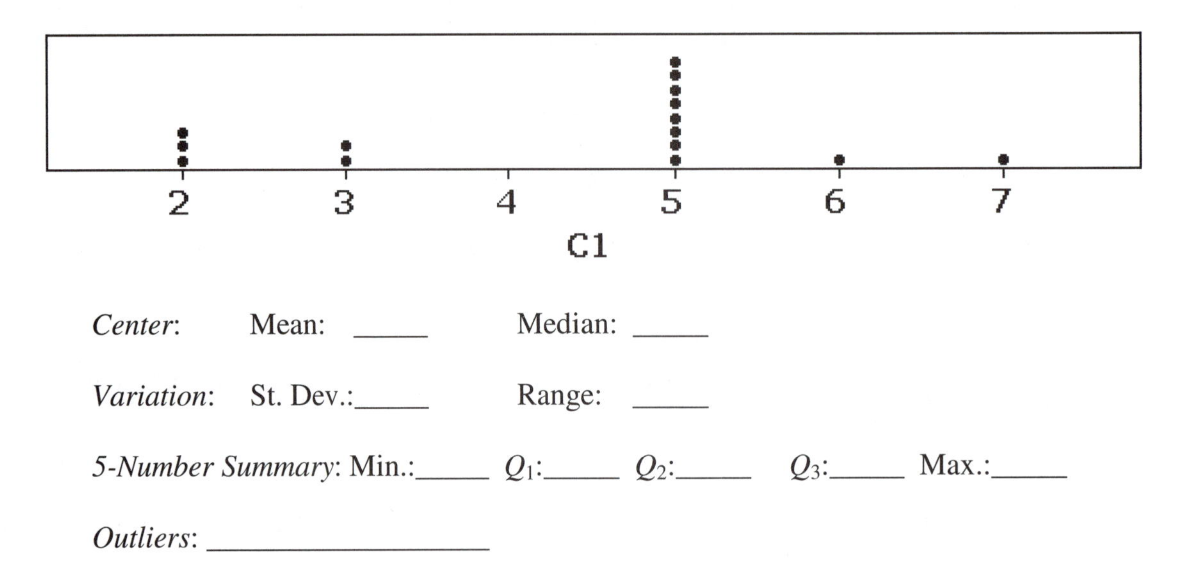

Center: Mean: _____ Median: _____

Variation: St. Dev.:_____ Range: _____

5-Number Summary: Min.:_____ Q_1:_____ Q_2:_____ Q_3:_____ Max.:_____

Outliers: __________________

3–12. ***Comparing Data*** Open the Minitab worksheet COLA and use Minitab to compare the weights of regular Coke and the weights of diet Coke. Obtain printouts of relevant results. What do you conclude? Can you explain any substantial difference?

____

____

3–13. ***Comparing Data*** In "Tobacco and Alcohol Use in G–Rated Children's Animated Films," by Goldstein, Sobel and Newman (*Journal of the American Medical Association,* Vol. 281, No. 12), the lengths (in seconds) of scenes showing tobacco use and alcohol use were recorded for animated children's movies. Refer to Data Set 8 in Appendix B from the textbook. (The Minitab worksheet is CHMOVIE.MTW.) Find the mean and median for the tobacco times, then find the mean and median for the alcohol times. Does there appear to be a difference between those times? Which appears to be the larger problem: scenes showing tobacco use or scenes showing alcohol use?

____

____

3–14. ***Working with Your Own Data*** Through observation or experimentation, collect your own set of sample values. Obtain at least 40 values and try to select data from an interesting population. List the values below.

____

____

____

____

____

____

____

____

Find the following results.

Center: Mean: _____ Median: _____

Variation: St. Dev.:_____ Range: _____

5-Number Summary: Min.:_____ Q_1:_____ Q_2:_____ Q_3:_____ Max.:_____

Outliers: ____________________

Use Minitab to further explore the data. Obtain printouts of relevant results. Describe the nature of the data. That is, what do the values represent? Describe important characteristics of the data set, and include Minitab printouts to support your observations.

4

Probabilities Through Simulations

4-1 Simulation Methods

Chapter 4 in the Triola textbook focuses on principles of probability theory. That chapter presents a variety of rules and methods for finding probabilities of different events. The textbook focuses on traditional approaches to computing probability values, but this chapter in this manual/workbook focuses instead on an alternative approach based on *simulations*.

A **simulation** of a procedure is a process that behaves the same way as the procedure, so that similar results are produced.

Mathematician Stanislaw Ulam once studied the problem of finding the probability of winning a game of solitaire, but the theoretical computations involved were too complicated. Instead, Ulam took the approach of programming a computer to simulate or "play" solitaire hundreds of times. The ratio of wins to total games played is the approximate probability he sought. This same type of reasoning was used to solve important problems that arose during World War II. There was a need to determine how far neutrons would penetrate different materials, and the method of solution required that the computer make various random selections in much the same way that it can randomly select the outcome of the rolling of a pair of dice. This neutron diffusion project was named the Monte Carlo Project and we now refer to general methods of simulating experiments as *Monte Carlo methods*. Such methods are the focus of this chapter. The concept of simulation is quite easy to understand with simple examples.

- We could simulate the rolling of a die by using Minitab to randomly generate whole numbers between 1 and 6 inclusive, provided that the computer selects from the numbers 1, 2, 3, 4, 5, and 6 in such a way that those outcomes are equally likely.
- We could simulate births by flipping a coin, where "heads" represents a baby girl and "tails" represents a baby boy. We could also simulate births by using Minitab to randomly generate 1s (for baby girls) and 0s (for baby boys).

It is extremely important to construct a simulation so that it behaves just like the real procedure. The following example illustrates a right way and a wrong way.

EXAMPLE Describe a procedure for simulating the rolling of a pair of dice.

SOLUTION In the procedure of rolling a pair of dice, each of the two dice yields a number between 1 and 6 (inclusive) and those two numbers are then added. Any simulation should do exactly the same thing.

Right way to simulate rolling two dice: Randomly generate one number between 1 and 6, then randomly generate another number between 1 and 6, then add the two results.

Wrong way to simulate rolling two dice: Randomly generate numbers between 2 and 12. This procedure is similar to rolling dice in the sense that the results are always between 2 and 12, but these outcomes between 2 and 12 are equally likely. With real dice, the values between 2 and 12 are *not* equally likely. This simulation would yield terrible results.

4-2 Minitab Simulation Tools

Minitab includes several different tools that can be used for simulations.

1. Click on the main menu item of **Calc**.
2. Click on the subdirectory item of **Random Data**.

You will get a long list of different options, but we will use four of them (integer, uniform, Bernouli, normal) that are particularly relevant to the simulation techniques of Chapter4 in the textbook.

- **Integer:** Generates a sample of randomly selected integers with a specified minimum and maximum. This Minitab tool is particularly good for a variety of different simulations. Shown below is the dialog box that is provided when you select **Calc/Random Data/ Integer**. The entries in this box tell Minitab to randomly generate 500 values between 1 and 6 inclusive, and put them in column C1. This simulates rolling a single die 500 times.

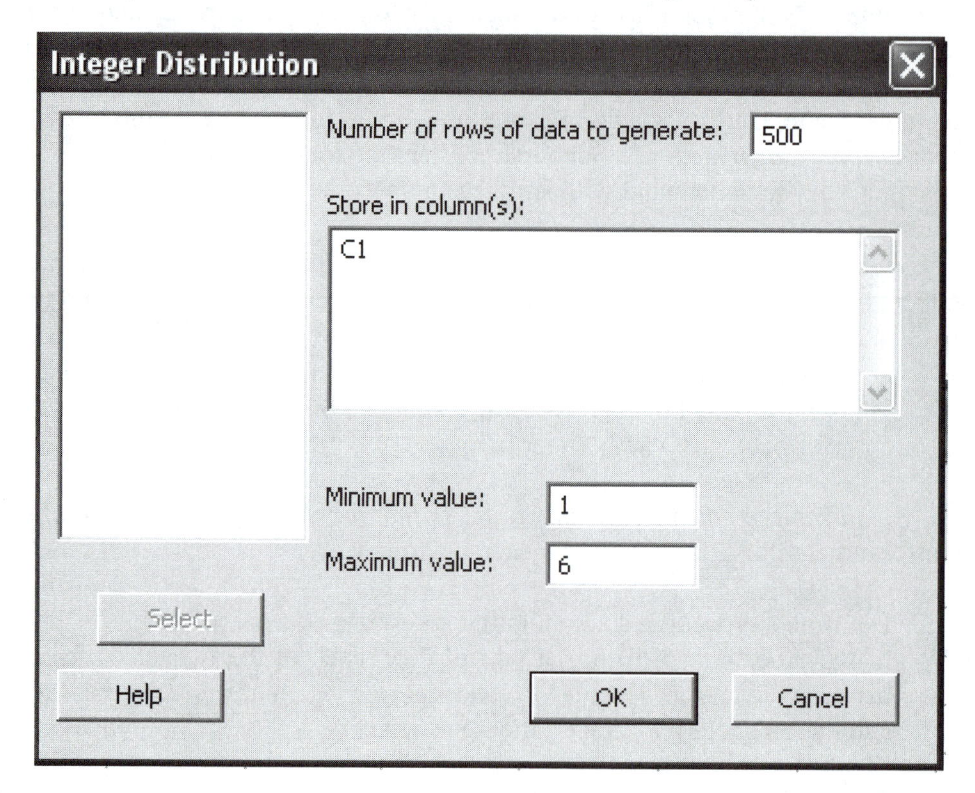

- **Uniform:** Select **Calc**, then **Random Data**, then **Uniform.** The generated values are "uniform" in the sense that all possible values have the same chance of being selected. The Minitab dialog box shown below has entries indicating that 500 values should be randomly selected between 0.0 and 9.0. If you want whole numbers, use the Integer distribution instead of the Uniform distribution.

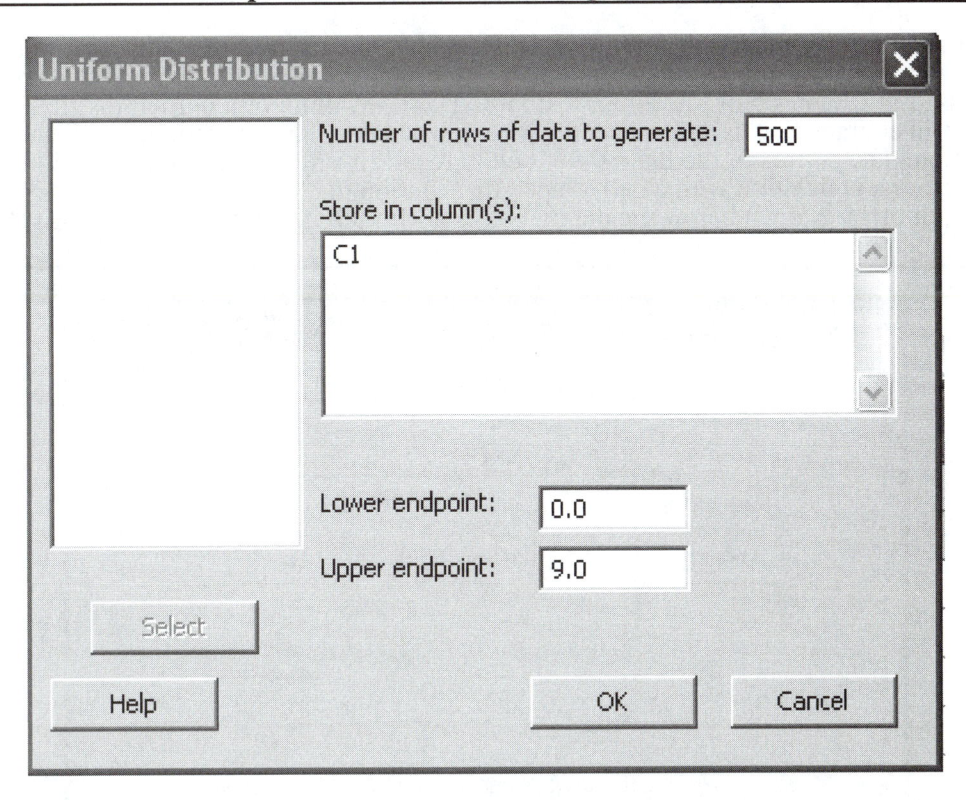

- **Bernouli:** Select **Calc**, then **Random Data**, then **Bernouli**. The dialog box below has entries that result in a column of 0s (failure) and 1s (success), where the probability of success is 0.5. This simulates flipping a coin 500 times. The entries can be changed for different circumstances.

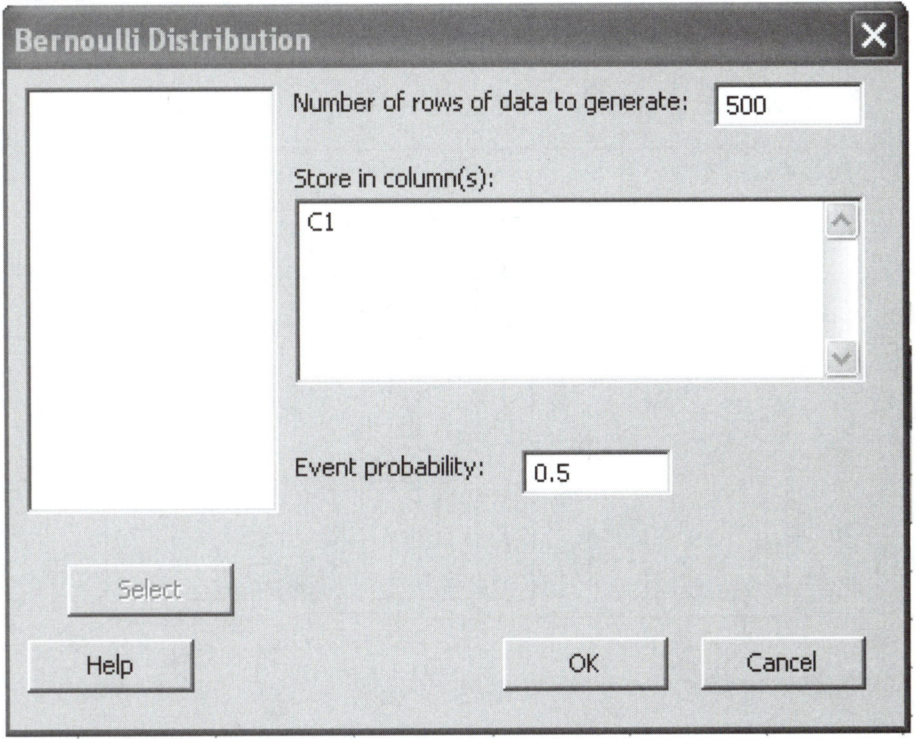

- **Normal:** Select **Calc**, then **Random Data**, then **Normal** to generate a sample of data randomly selected from a population having a normal distribution. (Normal distributions are discussed in Chapter 6 of the Triola textbook. For now, think of a normal distribution as one with a histogram that is bell–shaped.) You must enter the population mean and standard deviation. The entries in the dialog box below result in a simulated sample of 500 IQ scores taken from a population with a bell-shaped distribution that has a mean of 100 and a standard deviation of 15. Shown below the dialog box is a Minitab-generated histogram displaying the bell-shaped nature of the generated data.

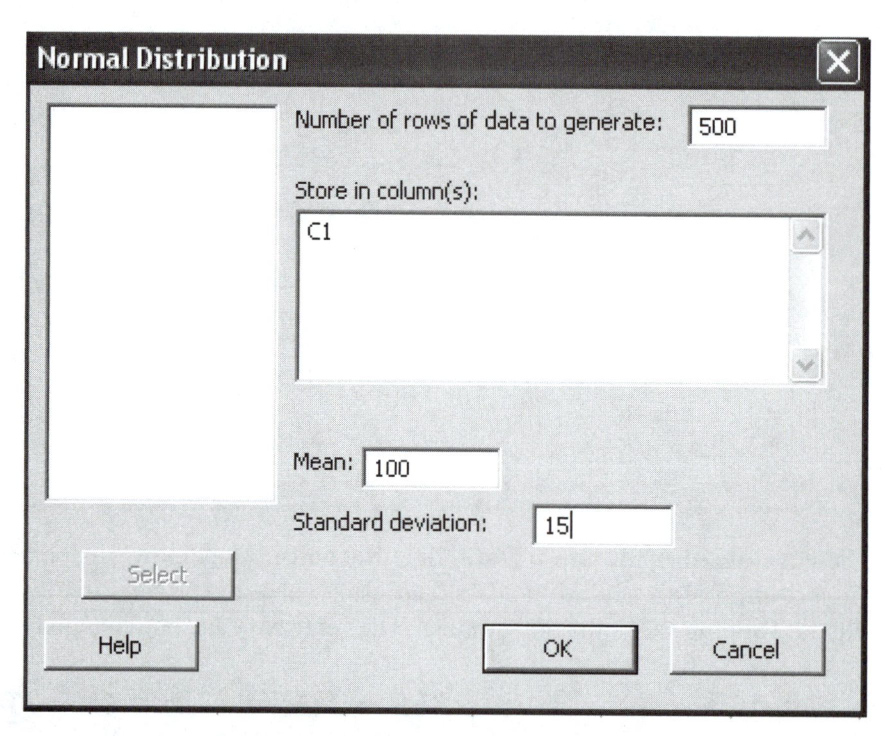

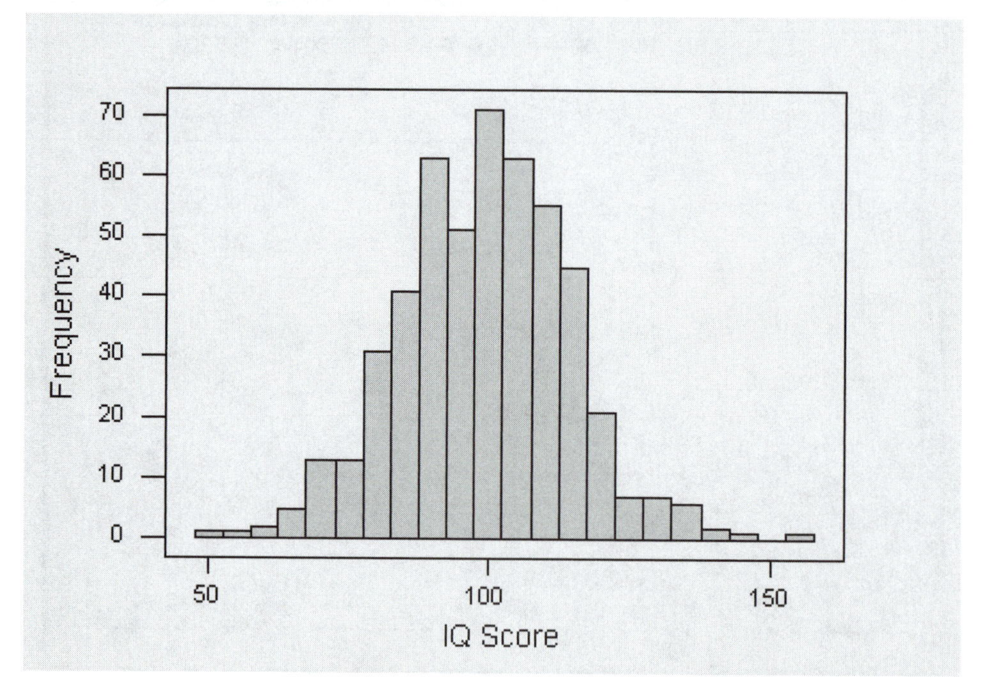

4-3 Simulation Examples

We will now illustrate the preceding Minitab features by describing specific simulations.

Simulation 1: Generating 50 births (boys/girls)

To simulate 50 births with the assumption that boys and girls are equally likely, use either of the following:

- Use Minitab's **Integer** distribution (see Section 4-2 of this manual/workbook) to generate 50 integers between 0 and 1. If you arrange the results in order (using the **Sort** feature as described in Section 3-4 of this manual/workbook), it is very easy to count the number of 0s (or boys) and the number of 1s (or girls).

- Use Minitab's **Bernouli** distribution. Enter 50 for the sample size and enter 0.5 for the probability. Again, it is very easy to count the number of 0s (or boys) and the number of 1s (or girls) if the data are sorted, as described in Section 3-4 of this manual/workbook.

Simulation 2: Rolling a *single* die 60 times

To simulate 60 rolls of a single die, use Minitab's **Integer** distribution to generate 60 integers between 1 and 6. Again, sorting the generated values makes it easy to count the number of 1s, 2s, and so on.

Simulation 3: Rolling a *pair* of dice 750 times

To simulate the rolling of a *pair* of dice 750 times, use column C1 for one die, use column C2 for the second die, then let column C3 be the *sum* of columns C1 and C2. That is, create column C1 by using the method described in Simulation 2 above. Then create column C2 by again using the method described in Simulation 2 above. Then use **Calc/Calculator** to enter the expression C1 + C2 with the result stored in column C3.

Simulation 4: Generating 25 birth dates

Instead of generating 25 results such as "March 5," or "November 27," randomly generate 25 integers between 1 and 365 inclusive. (We are ignoring leap years). Use Minitab's **Integer** distribution and enter 25 for the sample size. Also enter 1 for the minimum and enter 365 for the maximum. If you sort the simulated birth dates by using the procedure in Section 3-4 of this manual/workbook, it becomes easy to scan the sorted list and determine whether there are two birth dates that are the same. If there are two birth dates that are the same, they will show up as *consecutive* equal values in the sorted list.

CHAPTER 4 EXPERIMENTS: Probabilities Through Simulations

4–1. ***Birth Simulation*** Use Minitab to simulate 500 births, where each birth results in a boy or girl. Sort the results, count the number of girls, and enter that value here:_________

Based on that result, estimate the probability of getting a girl when a baby is born. Enter the estimated probability here: __________

The preceding estimated probability is likely to be different from 0.5. Does this suggest that the computer's random number generator is defective? Why or why not?

____

____

4–2. ***Dice Simulation*** Use Minitab to simulate 1000 rolls of a pair of dice. Sort the results, then find the number of times that the total was exactly 7. Enter that value here:_____

Based on that result, estimate the probability of getting a 7 when two dice are rolled. Enter the estimated probability here:__________

How does this estimated probability compare to the computed (theoretical) probability of 1/6 or 0.167? ____

____

4–3. ***Probability of at Least 11 Girls***

a. Use Minitab to simulate 20 births. Does the result consist of at least 11 girls? _____
b. Repeat part (a) nine more times and record the result from part (a) along with the other nine results here: ___ ___ ___ ___ ___ ___ ___ ___ ___ ___
c. Based on the results from part (b), what is the estimated probability of getting at least 11 girls in 20 births? __________

4-4. ***Brand Recognition*** The probability of randomly selecting an adult who recognizes the brand name of McDonald's is 0.95 (based on data from Franchise Advantage). Use Minitab to develop a simulation so that each individual outcome should be an indication of one of two results: (1) The consumer recognizes the brand name of McDonald's; (2) the consumer does not recognize the brand name of McDonald's. (*Hint:* Randomly generate integers between 1 and 100, and consider results from 1 through 95 to be people who recognize the brand name of McDonald's.) Conduct the simulation 1000 times and record the number of consumers who recognize the brand name of McDonald's. If possible, obtain a printed copy of the results. Is the proportion of those who recognize McDonald's reasonably close to the value of 0.95?

____

____

4-5. ***Simulating Hybridization*** When Mendel conducted his famous hybridization experiments, he used peas with green pods and yellow pods. One experiment involved crossing peas in such a way that 75% of the offspring peas were expected to have green pods, and 25% of the offspring peas were expected to have yellow pods. Use Minitab to simulate 1000 peas in such a hybridization experiment. Each of the 1000 individual outcomes should be an indication of one of two results: (1) The pod is green; (2) the pod is yellow. Is the percentage of yellow peas from the simulation reasonably close to the value of 25%?

____

____

4–6. ***Probability of at Least 55 Girls*** Use Minitab to conduct a simulation for estimating the probability of getting at least 55 girls in 100 births. Enter the estimated probability here:__________ Describe the procedure used to obtain the estimated probability.

____

In testing a gender-selection method, assume that the Biogene Technology Corporation conducted an experiment with 100 couples who were treated, and that the 100 births included at least 55 girls. What should you conclude about the effectiveness of the treatment?

____

4–7. ***Probability of at Least 65 Girls*** Use Minitab to conduct a simulation for estimating the probability of getting at least 65 girls in 100 births. Enter the estimated probability here:__________

Describe the procedure used to obtain the estimated probability.

____

In testing a gender-selection method, if the Biogene Technology Corporation conducted an experiment with 100 couples who were treated, and the 100 births included at least 65 girls, what should you conclude about the effectiveness of the treatment?

____

4-8. ***Gender-Selection Method*** As of this writing, the latest results available from the Microsort YSORT method of gender-selection consist of 127 boys in 152 births. That is, among 152 sets of parents using the YSORT method for increasing the likelihood of a boy, 127 actually had boys and the other 25 had girls. Assuming that the YSORT method has no effect and that boys and girls are equally likely, use Minitab to simulate 152 births. Is it unusual to get 127 boys in 152 births? What does the result suggest about the YSORT method?

____

____

4–9. ***Simulating Three Dice*** Develop a simulation for rolling three dice. Simulate the rolling of the three dice 100 times. Describe the simulation, then use it to estimate the probability of getting a total of 10 when three dice are rolled.

____

____

4–10. ***Simulating Left–Handedness*** Ten percent of us are left-handed. In a study of dexterity, people are randomly selected in groups of five. Develop a simulation for finding the probability of getting at least one left-handed person in a group of five. Simulate 100 groups of five. How does the probability compare to the correct result of 0.410, which can be found by using the probability rules in the textbook?

____

____

4–11. ***Nasonex Treatment*** Nasonex is a nasal spray used to treat allergies. In clinical trials, 1671 subjects were given a placebo, and 2 of them developed upper respiratory tract infections. Another 2103 patients were treated with Nasonex and 6 of them developed upper respiratory tract infections. Assume that Nasonex has no effect on upper respiratory tract infections, so that the rate of those infections also applies to Nasonex users. Using the placebo rate of 2/1671, simulate groups of 2103 subjects given the Nasonex treatment, then determine whether a result of 6 upper respiratory tract infections could easily occur. Describe the results. What do the results suggest about Nasonex as a cause of upper respiratory tract infections?

____

____

4–12. ***Birthdays*** Simulate a class of 25 birth dates by randomly generating 25 integers between 1 and 365. (We will ignore leap years.) Arrange the birth dates in ascending order, then examine the list to determine whether at least two birth dates are the same. (This is easy to do, because any two equal integers must be next to each other.)

Generated "birth dates:" ___ ___ ___ ___ ___ ___ ___ ___ ___ ___ ___ ___ ___
___ ___ ___ ___ ___ ___ ___ ___ ___ ___ ___ ___

Are at least two of the "birth dates" the same? __________

4—13. ***Birthdays*** Repeat the preceding experiment nine additional times and record all ten of the yes/no responses here:

____ ____ ____ ____ ____ ____ ____ ____ ____ ____

Based on these results, what is the probability of getting at least two birth dates that are the same (when a class of 25 students is randomly selected)? __________

4—14. ***Birthdays*** Repeat Experiments 4-12 and 4-13 for 50 people instead of 25. Based on the results, what is the estimated probability of getting at least two birth dates that are the same (when a class of 50 students is randomly selected)? __________

4—15. ***Birthdays*** Repeat Experiments 4-12 and 4-13 for 100 people instead of 25. Based on the results, what is the estimated probability of getting at least two birth dates that are the same (when a class of 100 students is randomly selected)? __________

4—16. ***Normally Distributed IQ Scores*** IQ scores are normally distributed with a mean of 100 and a standard deviation of 15. Generate a normally distributed sample of 800 IQ scores by using the given mean and standard deviation. Sort the results (arrange them in ascending order).

a. Examine the sorted results to estimate the probability of randomly selecting someone with an IQ score between 90 and 110 inclusive. Enter the result here.______
b. Examine the sorted results to estimate the probability of randomly selecting someone with an IQ score greater than 115. __________
c. Examine the sorted results to estimate the probability of randomly selecting someone with an IQ score less than 120. __________
d. Repeat part a of this experiment nine more times and list all ten probabilities here.

_____ _____ _____ _____ _____ _____ _____ _____ _____ _____

e. Examine the ten probabilities obtained above and comment on the *consistency* of the results.

____

f. How might we modify this experiment so that the results can become more consistent?

____

g. If the results appear to be very consistent, what does that imply about any individual sample result?

____

4—17. ***Law of Large Numbers*** In this experiment we test the Law of Large Numbers, which states that "as an experiment is repeated again and again, the empirical probability of success tends to approach the actual probability." We will use a simulation of a single die, and we will consider a success to be the outcome of 1. (Based on the classical definition of probability, we know that $P(1) = 1/6 = 0.167$.)

a. Simulate 5 trials by generating 5 integers between 1 and 6. Count the number of 6s that occurred and divide that number by 5 to get the empirical probability.
Based on 5 trials, $P(1) =$ _____.

b. Repeat part (a) for 25 trials. Based on 25 trials, $P(1)=$ _____.

c. Repeat part (a) for 50 trials. Based on 50 trials, $P(1)=$ _____.

d. Repeat part (a) for 500 trials. Based on 500 trials, $P(1)=$ _____.

e. Repeat part (a) for 1000 trials. Based on 1000 trials, $P(1) =$ _____.

f. In your own words, generalize these results in a restatement of the Law of Large Numbers.

____

____

____

____

4-18. ***Detecting Fabricated Results*** Consider a class experiment in which some students actually flip a coin 200 times and record the results, while others make up their own results for 200 coin flips. It is easy to identify the fabricated results using this criterion: If there is a run of six heads or six tails, the results are real, but if there is no such run, the results are fabricated. This is based on the principle that when fabricating results, people almost never include a run of six or more heads or tails, but with 200 actual coin flips, there is a very high probability of getting such a run of at least six heads or tails. The calculation for the probability of getting a run of at least six heads or six tails is *extremely* difficult, so conduct simulations to estimate that probability. Describe the process and identify the estimated probability that when 200 coins are tossed, there is run of at least six heads or tails.

____

____

____

____

4–19. ***Sticky Probability Problem*** Consider the following exercise, which is extremely difficult to solve with the formal rules of probability.

Two points along a straight stick are randomly selected. The stick is then broken at these two points. Find the probability that the three pieces can be arranged to form a triangle.

Instead of attempting a solution using formal rules of probability, we will use a simulation. The length of the stick is irrelevant, so assume it's one unit long and its length is measured from 0 at one end to 1 at the other end. Use Minitab to randomly select the two break points with the random generation of two numbers from a uniform distribution with a minimum of 0, a maximum of 1, and 4 decimal places. Plot the break points on the "stick" below.

0____1

A triangle can be formed if the longest segment is less than 0.5, so enter the lengths of the three pieces here: __________ __________ __________

Can a triangle be formed?

Now repeat this process nine more times and summarize all of the results below.

Trial	Break Points		Triangle formed?
1			
2			
3			
4			
5			
6			
7			
8			
9			
10			

Based on the ten trials, what is the estimated probability that a triangle can be formed? ______ This estimate gets better with more trials.

5

Probability Distributions

5-1 Exploring Probability Distributions

Chapter 5 in the Triola textbook discusses probability distributions, and its focus is *discrete* probability distributions only. These important definitions are introduced:

Definitions

A **random variable** is a variable (typically represented by x) that has a single numerical value, determined by chance, for each outcome of a procedure.

A **probability distribution** is a description that gives the probability for each value of the random variable. It is often expressed in the format of a table, formula, or graph.

When working with a probability distribution, we should consider the same important characteristics introduced in Chapter 2:

1. **Center:** Measure of center, which is a representative or average value that gives us an indication of where the middle of the data set is located

2. **Variation:** A measure of the amount that the values vary among themselves

3. **Distribution:** The nature or shape of the distribution of the data, such as bell-shaped, uniform, or skewed

4. **Outliers:** Sample values that are very far away from the vast majority of the other sample values

5. **Time:** Changing characteristics of the data over time

Section 5-1 of this manual/workbook addresses these important characteristics for probability distributions. The characteristics of center and variation are addressed with formulas for finding the mean, standard deviation, and variance of a probability distribution. The characteristic of distribution is addressed through the graph of a probability histogram.

Table 5-1 Although Minitab is not designed to deal directly with a probability distribution, it can often be used. Consider Table 5-1 from the textbook, reproduced here. This table lists the probabilities for the number girls in 2 births.

Table 5-1

Number of Girls in 2 Births

x	$P(x)$
0	0.25
1	0.50
2	0.25

If you examine the data in the table, you can verify that a probability distribution is defined because the three key requirements are satisfied:

1. The variable x is a numerical random variable and its values are associated with probabilities, as in Table 5-1.
2. The sum of the probabilities is 1, as required. (0.25 + 0.50 + 0.25 = 1.)
3. Each value of $P(x)$ is between 0 and 1. (Specifically, 0.25 and 0.50 and 0.25 are each between 0 and 1 inclusive.)

Having determined that Table 5-1 does define a probability distribution, let's now see how we can use Minitab to find the mean μ and standard deviation σ. Minitab is not designed to directly calculate the value of the mean and standard deviation, but there is a way to those values, although it is not easy.

Finding μ and σ for a Probability Distribution

1. Enter the values of the random variable x in column C1.
2. Enter the corresponding probabilities in column C2.
3. Calculate the mean μ by using Formula 5-1 in the textbook for the mean of a probability distribution: $\mu = \sum[x \cdot P(x)]$. This is accomplished by clicking on **Calc**, selecting **Calculator**, and entering the expression sum(C1*C2) with the result stored in column C3, as shown in the following screen.

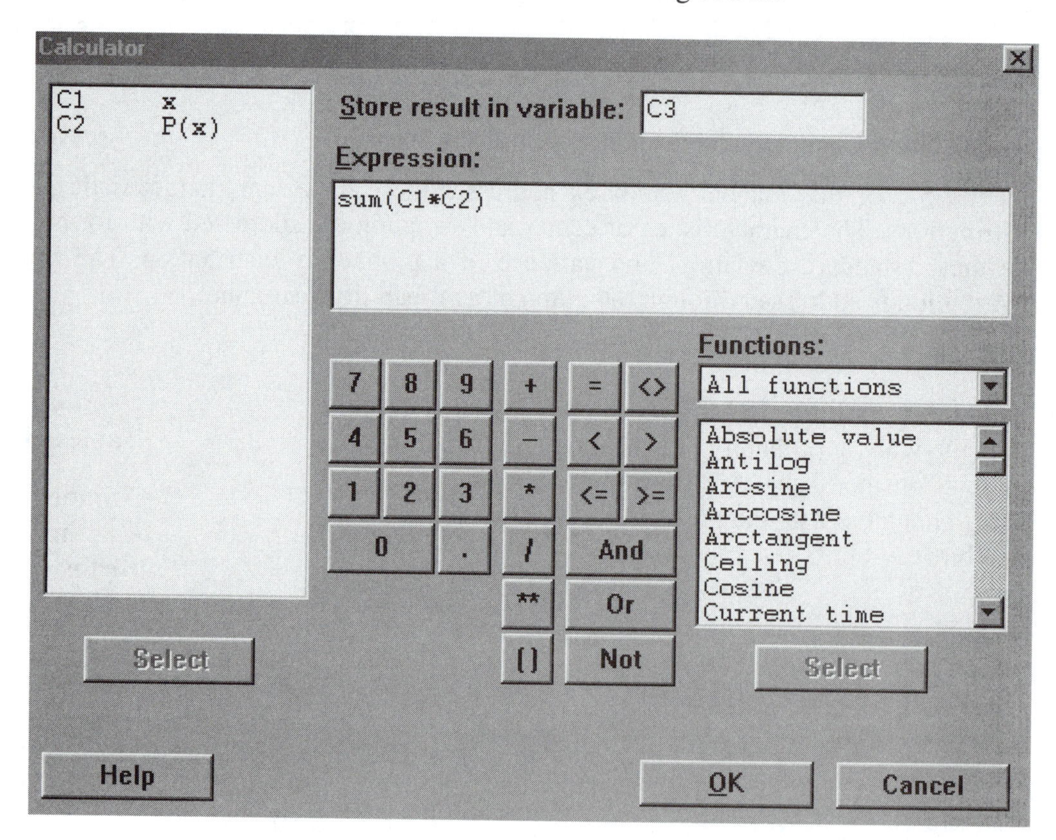

4. Now proceed to calculate the value of the standard deviation σ by using Formula 5-4 from the textbook: $\sigma = \sqrt{\left[\Sigma x^2 \cdot P(x)\right] - \mu^2}$. This is accomplished by selecting **Calc**, then **Calculator**, and entering the expression

sqrt(sum(C1**2*C2) – C3**2)

with the result stored in column C4.

If you use the above procedure with Table 5-1, the result will be as shown below. We can see that $\mu = 1$ and the standard deviation is $\sigma = 0.707107$ (or 0.7 after rounding). The names of the columns in the display below were manually entered.

x	P(x)	Mean	St. Dev.
0	0.25	1	0.707107
1	0.50		
2	0.25		

5-2 Binomial Distributions

A **binomial distribution** is defined in Section 5-3 of the Triola textbook to be to be a probability distribution that meets all of the following requirements:

1. The experiment must have a fixed number of trials.
2. The trials must be independent. (The outcome of any individual trial doesn't affect the probabilities in the other trials.)
3. Each trial must have all outcomes classified into two categories.
4. The probabilities must remain constant for each trial.

We also introduced notation with S and F denoting success and failure for the two possible categories of all outcomes. Also, p and q denote the probabilities of S and F, respectively, so that $P(S) = p$ and $P(F) = q$. We also use the following symbols.

n denotes the fixed number of trials

x denotes a specific number of successes in n trials so that x can be any whole number between 0 and n, inclusive.

p denotes the probability of success in *one* of the n trials.

q denotes the probability of failure in *one* of the n trials.

$P(x)$ denotes the probability of getting exactly x successes among the n trials.

Section 5-3 in the Triola textbook describes three methods for determining probabilities in binomial experiments. Method 1 uses the binomial probability formula:

$$P(x) = \frac{n!}{(n-x)!x!} p^x q^{n-x}$$

Method 2 requires computer usage. We noted in the textbook that if a computer and software are available, this method of finding binomial probabilities is fast and easy, as shown in the Minitab procedure that follows. We illustrate the Minitab procedure with the following problem:

> **EXAMPLE** ***Devil of a Problem*** Based on a recent Harris poll, 60% of adults believe in the devil. Assuming that we randomly select 5 adults, find the probability that exactly 3 of the 5 adults believe in the devil

For this example, $n = 5$, $p = 0.60$, and the possible values of x are 0, 1, 2, 3, 4, 5.

Minitab Procedure for Finding Probabilities with a Binomial Distribution

1. In column C1, enter the values of the random variable x for which you want probabilities. (You can enter all values of x as 0, 1, 2, . . .) Here is the entry for the x values of 0, 1, 2, 3, 4, 5:

x
0
1
2
3
4
5

2. Click on **Calc** from the main menu.
3. Select **Probability Distributions**.
4. Select **Binomial**.
5. The entries in the dialog box shown below correspond to a binomial distribution with $n = 5$, $p = 0.60$ (the probability of success for one trial). The probability values are to be computed for the values of x that have been entered in column C1, and the results will be stored in column C2.

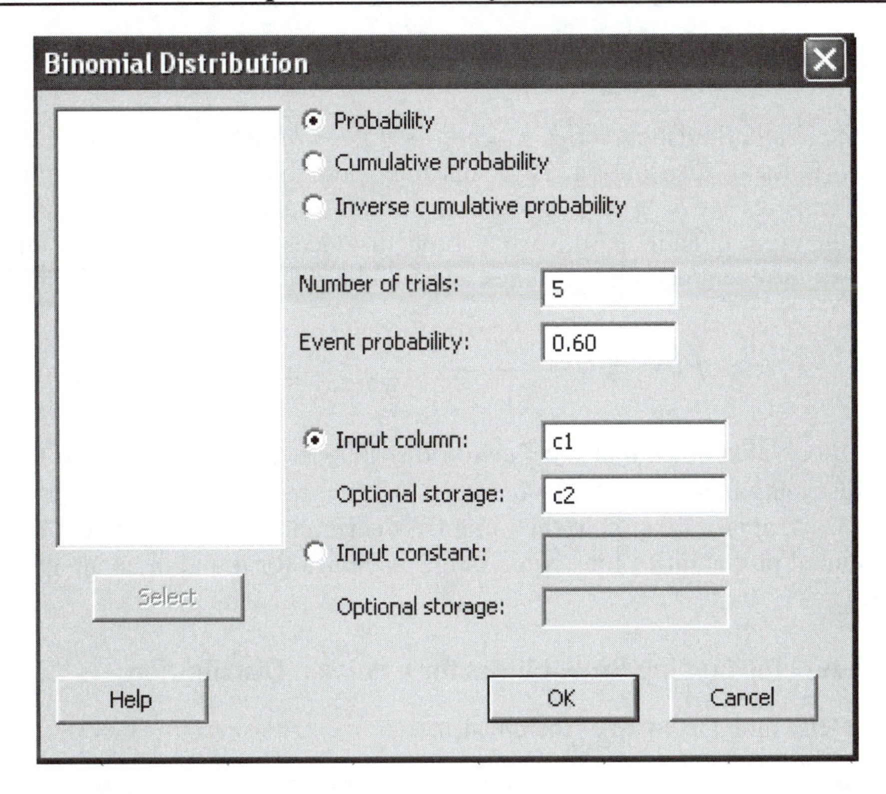

After clicking on **OK**, the probabilities will be displayed in column C2, as shown below. (The column heading of $P(x)$ was manually entered.)

x	P(x)
0	0.01024
1	0.07680
2	0.23040
3	0.34560
4	0.25920
5	0.07776

From this display, we see that $P(0) = 0.01024$, $P(1) = 0.07680$, and so on. The preceding example requires the probability that exactly 3 of 5 adults believe in the devil, and the result is $P(3) = 0.346$ (rounded).

5-3 Poisson Distributions

Textbooks in the Triola Statistics Series (except for *Essentials of Statistics*) discuss the Poisson distribution. A Poisson distribution is a discrete probability distribution that applies to occurrences of some event *over a specified interval*. The random variable x is the number of occurrences of the event in an interval, such as time, distance, area, volume, or some similar unit. The probability of the event occurring x times over an interval is given by this formula:

$$P(x) = \frac{\mu^x \cdot e^{-\mu}}{x!} \quad \text{where } e \approx 2.71828$$

The textbook also notes that the Poisson distribution is sometimes used to approximate the binomial distribution when $n \geq 100$ and $np \leq 10$; in such cases, we use $\mu = np$. If using Minitab, the Poisson approximation to the binomial distribution isn't used much, because we can easily find binomial probabilities for a wide range of values for n and p, so an approximation is not necessary.

Minitab Procedure for Finding Probabilities for a Poisson Distribution

1. Determine the value of the mean μ.
2. In column C1, enter the values of the random variable x for which you want probabilities. (You can enter values of x as 0, 1, 2, ...)
3. Click on **Calc** from the main menu.
4. Select **Probability Distributions**.
5. Select **Poisson**.
6. You will now see a dialog box like the one shown below. The probability values are to be computed for the values of x that have been entered in column C1, and the results will be stored in column C2.

Let's consider Atlantic hurricanes occurring so that the mean number of hurricanes is 5.3 per year. If we want the probabilities corresponding to $x = 0, 1, 2, 3, 4, 5$ hurricanes in a year, enter the values of 0, 1, 2, 3, 4, 5 in column C1. The resulting probabilities are shown on the next page.

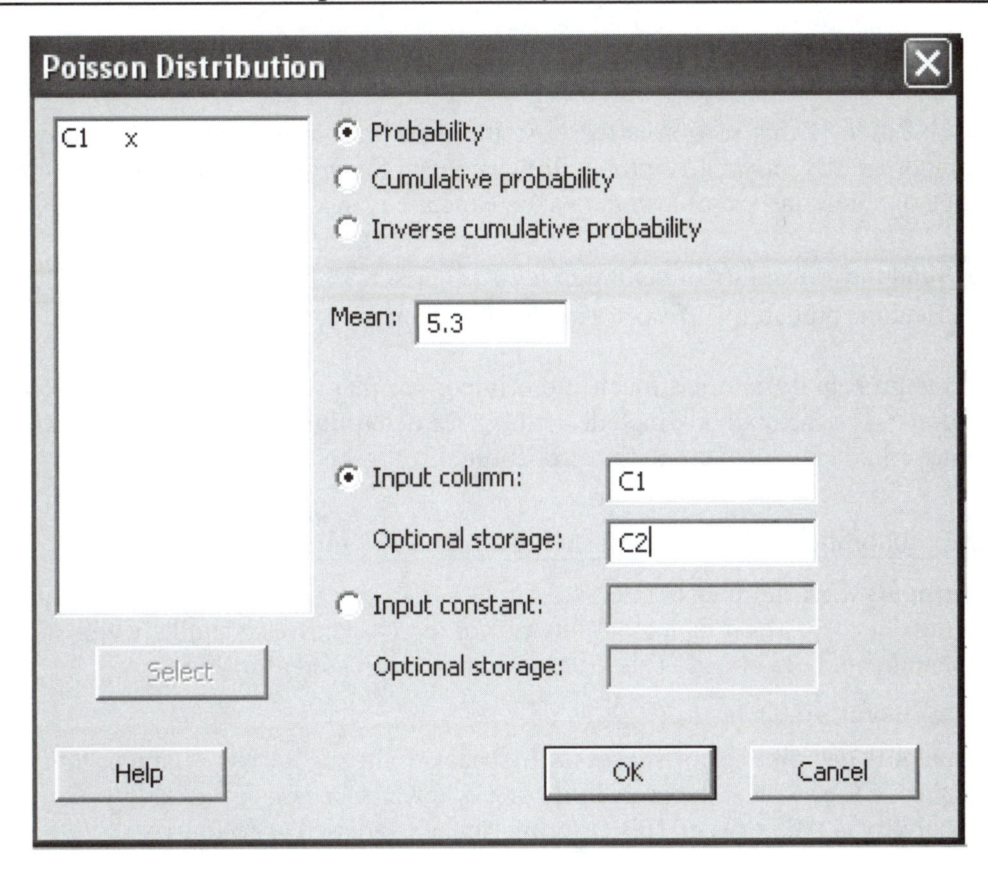

This is the Minitab result showing the probabilities calculated using the Poisson distribution. Additional values of *x* could have been entered, and the corresponding probabilities would have been found.

x	P(x)
0	0.004992
1	0.026455
2	0.070107
3	0.123856
4	0.164109
5	0.173955

5-4 Cumulative Probabilities

The main objective of this section is to stress the point that *cumulative probabilities* are often critically important. By *cumulative* probability, we mean the probability that the random variable x has a range of values instead of a single value. Here are typical examples:

- Find the probability of getting *at least* 13 girls in 14 births.
- Find the probability of *more than* 5 wins when roulette is played 200 times.

A *cumulative* probability is often much more important than the probability of any individual event. Section 5-2 of the textbook includes criteria for determining when results are *unusual*, and the following criteria involve cumulative probabilities.

Using Probabilities to Determine When Results Are Unusual

- **Unusually *high* number of successes:** x successes among n trials is an *unusually high* number of successes if the probability of x or more successes is unlikely with a probability of 0.05 or less. This criterion can be expressed as follows:

  $P(x \text{ or more}) \leq 0.05$.*
- **Unusually *low* number of successes:** x successes among n trials is an *unusually low* number of successes if the probability of x or fewer successes is unlikely with a probability of 0.05 or less. This criterion can be expressed as follows:

  $P(x \text{ or fewer}) \leq 0.05$.*

  *The value 0.05 is not absolutely rigid. Other values, such as 0.01, could be used to distinguish between results that can easily occur by chance and events that are very unlikely to occur by chance.

Example 7 in Section 5-2 of the textbook includes this: The Chapter Problem includes results consisting of 879 girls in 945 births. Is 879 girls in 945 births an unusually high number of girls? What does it suggest about the effectiveness of the XSORT method of gender selection?

Here we want the cumulative probability of getting 879 or more girls, assuming that boys and girls are equally likely. Using Minitab with the methods in Section 5-2 of this manual/workbook, we find that the probability of 879 or more girls is 0.000 when rounded to three decimal places. This indicates that 879 girls in 945 is an unusually high number of girls, suggesting that the XSORT method of gender selection is effective. Here, cumulative probabilities play a critical role in identifying results that are considered to be *unusual*. Later chapters focus on this important concept. If you examine the dialog boxes used for finding binomial probabilities and Poisson probabilities, you can see that in addition to obtaining probabilities for specific numbers of successes, you can also obtain *cumulative* probabilities.

CHAPTER 5 EXPERIMENTS: Probability Distributions

5-1. ***Genetic Disorder*** Three males with an X–linked genetic disorder have one child each. The random variable x is the number of children among the three who inherit the X–linked genetic disorder, and the probability distribution for that number of children is given in the table below.

x	$P(x)$
0	0.4219
1	0.4219
2	0.1406
3	0.0156

Use Minitab with the procedure described in Section 5-1 of this manual/workbook for the following.

Find the mean and standard deviation:

Mean:______ St. Dev.:____________

5-2. ***Overbooked Flights.*** Air America has a policy of routinely overbooking flights. The random variable x represents the number of passengers who cannot be boarded because there are more passengers than seats (based on data from an IBM research paper by Lawrence, Hong, and Cherrier).

x	$P(x)$
0	0.051
1	0.141
2	0.274
3	0.331
4	0.187

Use Minitab with the procedure described in Section 5-1 of this manual/workbook for the following.

Find the mean and standard deviation:

Mean:______ St. Dev.:____________

5-3. ***Genetics.*** Groups of five babies are randomly selected. In each group, the random variable x is the number of babies with green eyes (base on data from a study by Dr. Sorita Soni at Indiana University). (The symbol 0+ denotes a positive probability value that is very small.)

x	$P(x)$
0	0.528
1	0.360
2	0.098
3	0.013
4	0.001
5	0+

Use Minitab with the procedure described in Section 5-1 of this manual/workbook for the following.

Find the mean and standard deviation:

Mean:______ St. Dev.:_____________

5-4. ***Binomial Probabilities*** Use Minitab to find the binomial probabilities corresponding to $n = 4$ and $p = 0.05$. Enter the results below, along with the corresponding results found in Table A-1 of the textbook.

x	$P(x)$ from Minitab	$P(x)$ from Table A-1
0		
1		
2		
3		
4		

By comparing the above results, what advantage does Minitab have over Table A-1?

____

____

5-5. ***Binomial Probabilities*** Consider the binomial probabilities corresponding to $n = 4$ and $p = 1/4$ (or 0.25). If you attempt to use Table A-1 for finding the probabilities, you will find that the table does not apply to a probability of $p = 1/4$. Use Minitab to find the probabilities and enter the results below.

$P(0) =$ ______________

$P(1) =$ ______________

$P(2) =$ ______________

$P(3) =$ ______________

$P(4) =$ ______________

5-6. ***Binomial Probabilities*** Assume that boys and girls are equally likely and 100 births are randomly selected. Use Minitab with $n = 100$ and $p = 0.5$ to find $P(x)$, where x represents the number of girls among the 100 babies.

$P(35) =$ ______________

$P(45) =$ ______________

$P(50) =$ ______________

5-7. ***Cumulative Probabilities*** Assume that $P(\text{boy}) = 0.5121$, $P(\text{girl}) = 0.4879$, and that 100 births are randomly selected. Use Minitab to find the probability that the number of girls among 100 babies is . . .

a. Fewer than 60 ______________

b. Fewer than 48 ______________

c. At most 30 ______________

d. At least 55 ______________

e. More than 40 ______________

5-8. ***Identifying 0+*** In Table A-1 from the textbook, the probability corresponding to $n = 7$, $p = 0.95$, and $x = 2$ is shown as 0+. Use Minitab to find the corresponding probability and enter the result here. ______________

5-9. ***Identifying 0+*** In Table A-1 from the textbook, the probability corresponding to $n = 8$, $p = 0.01$, and $x = 6$ is shown as 0+. Use Minitab to find the corresponding probability and enter the result here. ______________

5-10. ***Identifying a Probability Distribution*** Use Minitab to construct a table of x and $P(x)$ values corresponding to a binomial distribution in which $n = 5$ and $p = 0.35$. Enter the table in the space below.

Table:

Binomial *Exercises 5*–11 through 5-15 involve binomial distributions. Use Minitab for those exercises.

5-11. ***Brand Recognition*** The brand name of Mrs. Fields (cookies) has a 90% recognition rate (based on data from Franchise Advantage). If Mrs. Fields herself wants to verify that rate by beginning with a small sample of ten randomly selected consumers, find the probability that exactly nine of the ten consumers recognize her brand name. Also find the probability that the number who recognize her brand name is *not* nine.

____

5-12. ***Too Young to Tat*** Based on a Harris poll, among adults who regret getting tattoos, 20% say that they were too young when they got their tattoos. Assume that 5 adults who regret getting tattoos are randomly selected, and find the indicated probability.

a. Find the probability that none of the selected adults say that they were too young to get tattoos. __________

b. Find the probability that exactly one of the selected adults says that they were too young to get tattoos. __________

c. Find the probability that the number of selected adults saying they were too young is 0 or 1. __________

5-13. ***Eye Color*** In the United States, 40% of the population have brown eyes (based on data from Dr. P Sorita Soni at Indiana University). If 14 people are randomly selected, find the probability that at least 12 of them have brown eyes. Is it unusual to randomly select 14 people and find that at least 12 of them have brown eyes? Why or why not?

____

5-14. ***Credit Rating*** There is a 1% delinquency rate for consumers with FICO (Fair Isaac & Company) credit rating scores above 800. If the Jefferson Valley Bank provides large loans to 12 people with FICO scores above 800, what is the probability that at least one of them becomes delinquent? Based on that probability, should the bank plan on dealing with a delinquency?

____

5-15. ***Genetics*** Ten peas are generated from parents having the green/yellow pair of genes, so there is a 0.75 probability that an individual pea will have a green pod. Find the probability that among the 10 offspring peas, at least nine have green pods. Is it unusual to get at least nine peas with green pods when ten offspring peas are generated? Why or why not?

____

Poisson *Exercises 5-16 through 5-18 involve Poisson distributions. Use Minitab for these exercises.*

5–16. ***Radioactive Decay*** Radioactive atoms are unstable because they have too much energy. When they release their extra energy, they are said to decay. When studying Cesium 137, it is found that during the course of decay over 365 days, 1,000,000 radioactive atoms are reduced to 977,287 radioactive atoms.

a. Find the mean number of radioactive atoms lost through decay in a day. ______

b. Find the probability that on a given day, 50 radioactive atoms decayed. ______

5-17. ***Chocolate Chip Cookies*** In the production of chocolate chip cookies, we can consider each cookie to be the specified interval unit required for a Poisson distribution, and we can consider the variable x to be the number of chocolate chips in a cookie. Table 3-1 is included with the Chapter Problem for Chapter 3 in the textbook, and it includes the numbers of chocolate chips in 34 different Keebler cookies. The Poisson distribution requires a value for μ, so use 30.4, which is the mean number of chocolate chips in the 34 Keebler cookies. Assume that the Poisson distribution applies.

a. Find the probability that a cookie will have 26 chocolate chips. __________

b. Find the probability that a cookie will have 30 chocolate chips. __________

5-18. ***Deaths From Horse Kicks*** A classic example of the Poisson distribution involves the number of deaths caused by horse kicks of men in the Prussian Army between 1875 and 1894. Data for 14 corps were combined for the 20-year period, and the 280 corps-years included a total of 196 deaths. After finding the mean number of deaths per corps-year, find the probability that a randomly selected corps-year has the following numbers of deaths.

a. 0 _____ b. 1 _____ c. 2 _____ d. 3 _____ e. 4 _____

The actual results consisted of these frequencies: 0 deaths (in 144 corps-years); 1 death (in 91 corps-years); 2 deaths (in 32 corps-years); 3 deaths (in 11 corps-years); 4 deaths (in 2 corps-years). Compare the actual results to those expected from the Poisson probabilities. Does the Poisson distribution serve as a good device for predicting the actual results?

____

____

6

Normal Distributions

6-1 Finding Areas and Values with a Normal Distribution

The Triola textbook describes methods for working with standard and nonstandard normal distributions. (A **standard normal distribution** has a mean of 0 and a standard deviation of 1.) Table A-2 in Appendix B of the textbook lists a wide variety of different z scores along with their corresponding areas. Minitab can also be used to find probabilities associated with the normal distribution. Although Table A-2 in the textbook includes only limited values of z scores and their corresponding areas, Minitab can be used with infinitely many different possible choices. In the following procedure, note that like Table A-2 in the textbook, Minitab areas are to the *left* of the related x value. Here is the procedure for finding areas similar to those in the body of Table A-2 from the textbook.

Minitab Procedure for Finding Areas and Values with a Normal Distribution

1. Select **Calc** from the main menu at the top of the screen.
2. Select **Probability Distributions** from the subdirectory.
3. Select **Normal**.
4. You will get a dialog box like the one shown below.

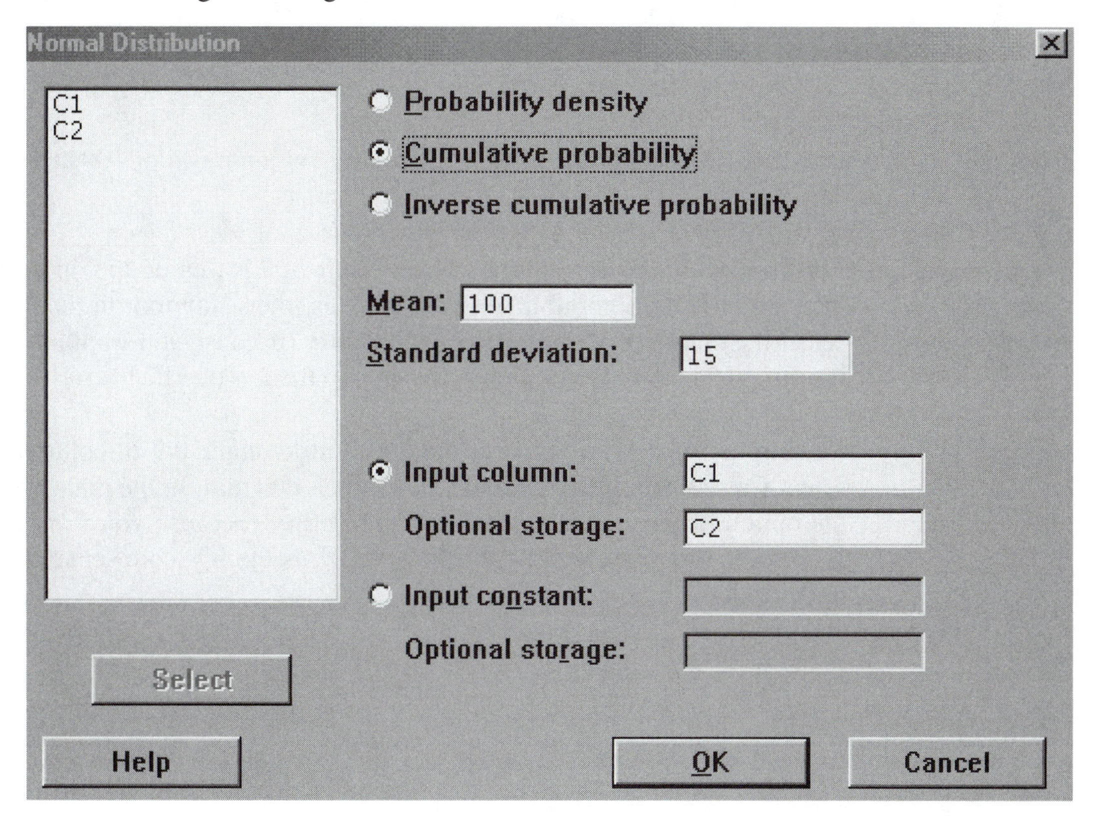

When making choices and entries in the dialog box, consider the following:

- **Probability density** is the height of the normal distribution curve, so we will rarely use this option.

- **Cumulative probability:** Use this option to find an area or probability. Note that you will get the area under the curve to the *left* of the x value(s) entered in column C1. (You must have already entered the x value(s) in column C1, or you could enter a constant value of x by selecting the **Input constant** option.)

- **Inverse cumulative probability:** Use this option to find a value corresponding to a given area. The resulting x value(s) separates an area to the *left* that is specified in column C1. (You must have already entered the cumulative left areas in column C1, or you could enter a single area as a constant by selecting the **Input constant** option.)

The input column is the column containing values (either areas or x values) that you have already entered, and the output column is the column that will list the results (either x values or areas).

5. Click **OK** and the results will be listed in the output column.

Examples: Assume that IQ scores are normally distributed with a mean of 100 and a standard deviation of 15.

- **Finding Area:** To find the area to the left of 115, enter 115 in column C1, then select **Calc**, **Probability Distributions**, then **Normal**. In the dialog box, select the option of **Cumulative Probability** (because you want an *area*). The result will be 0.843415, which is the area to the *left* of 115.

- **Finding a Value:** To find the 90th percentile, enter 0.9 in column C1, then select **Calc**, **Probability Distributions**, then **Normal**. In the dialog box, select the option of **Inverse Cumulative Probability** (because you want a *value* of IQ score). The result will be 119.223, which is the IQ score separating an area of 0.9 to its *left*.

6-2 Simulating and Generating Normal Data

We can learn much about the behavior of normal distributions by analyzing samples obtained from them. Sampling from real populations is often time consuming and expensive, but we can use the wonderful power of computers to obtain samples from theoretical normal distributions, and Minitab has such a capability, as described below. Let's consider IQ scores. IQ tests are designed to produce a mean of 100 and a standard deviation of 15, and we expect that such scores are normally distributed. Suppose we want to learn about the variation of sample means for samples of IQ scores. Instead of going out into the world and randomly selecting groups of people and administering IQ tests, we can sample from theoretical populations. We can then learn much about the distribution of sample means. The following procedure allows you to obtain a random sample from a normally distributed population with a given mean and standard deviation.

Generating a Random Sample from a Normally Distributed Population

1. Click on the main menu item of **Calc**.
2. Click on **Random Data**.
3. Click on **Normal**.
4. You will now get a dialog box such as the one shown below. You must enter the population mean and standard deviation. The entries in the dialog box below result in simulated sample of 500 IQ scores taken from a population with a normal distribution that has a mean of 100 and a standard deviation of 15. The sample values are stored in column C1.

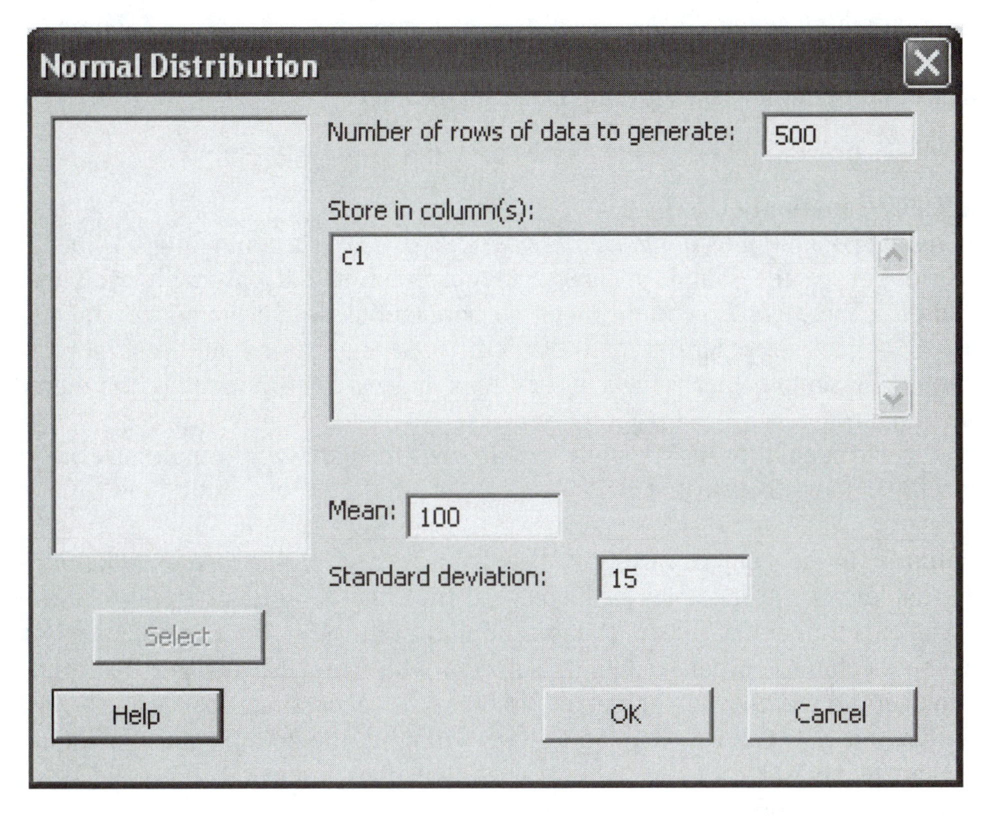

Important: *When using the above procedure, the generated sample will not have a mean equal to the specified population mean, and the standard deviation of the sample will not have a value equal to the specified standard deviation.* When the author generated 500 sample values with the above dialog box, he obtained sample values with a mean of 100.28 and a standard deviation of 13.98. Remember, you are sampling from a population with the specified parameters; you are not generating a sample with statistics exactly equal to the specified values. See Section 6–5 of this manual/workbook for a procedure that can be used to generate a sample with exact values of a specified mean and standard deviation.

6-3 The Central Limit Theorem

The textbook introduces the central limit theorem, which is then used in subsequent chapters when the important topics of estimating parameters and hypothesis testing are discussed. Here is the statement of the central limit theorem in the Triola textbook:

The Central Limit Theorem and the Sampling Distribution of $\bar{x}$

Given:

1. The random variable x has a distribution (which may or may not be normal) with mean μ and standard deviation σ.
2. Simple random samples all of the same size n are selected from the population. (The samples are selected so that all possible samples of size n have the same chance of being selected.)

Conclusions:

1. The distribution of sample means $\bar{x}$ will, as the sample size increases, approach a *normal* distribution.
2. The mean of all sample means is the population mean μ.
3. The standard deviation of all sample means is $\sigma/\sqrt{n}$.

Practical Rules Commonly Used

1. If the original population is *not normally distributed*, here is a common guideline: For $n > 30$, the distribution of the sample means can be approximated reasonably well by a normal distribution. (There are exceptions, such as populations with very nonnormal distributions requiring sample sizes larger than 30, but such exceptions are relatively rare.) The distribution of sample means gets closer to a normal distribution as the sample size n becomes larger.
2. If the original population is *normally distributed*, then for *any* sample size n, the sample means will be normally distributed.

Example In the study of methods of statistical analysis, it is extremely helpful to have a clear understanding of the statement of the central limit theorem. It is helpful to use Minitab in conducting an experiment that illustrates the central limit theorem. For example, let's use Minitab to generate five columns of data randomly selected with Minitab's **Integer** distribution, which generates integers so that they are all equally likely. After generating 5000 values (each between 0 and 9) in columns C1, C2, C3, C4, and C5, we will stack the 25000 values in column C6. We will also calculate the 5000 sample means of the values in columns C1, C2, C3, C4, and C5 and enter the results in column C7. (The sample means can be obtained by selecting **Calc,** then selecting **Row Statistics**; select **Mean** and enter C1-C7 for Input variables.) See the histograms shown below. The first histogram shows the distribution of the 25,000 generated integers, but the second histogram shows the distribution of the 5000 sample means. We can clearly see that the sample means have a distribution that is approximately normal, even thought the original population has a uniform distribution.

Distribution of 25,000 randomly selected integers

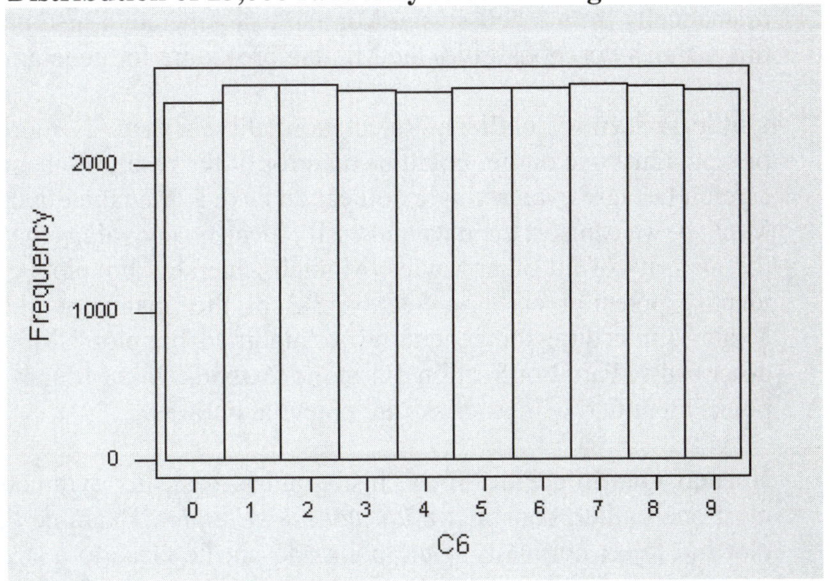

Distribution of 5000 sample means

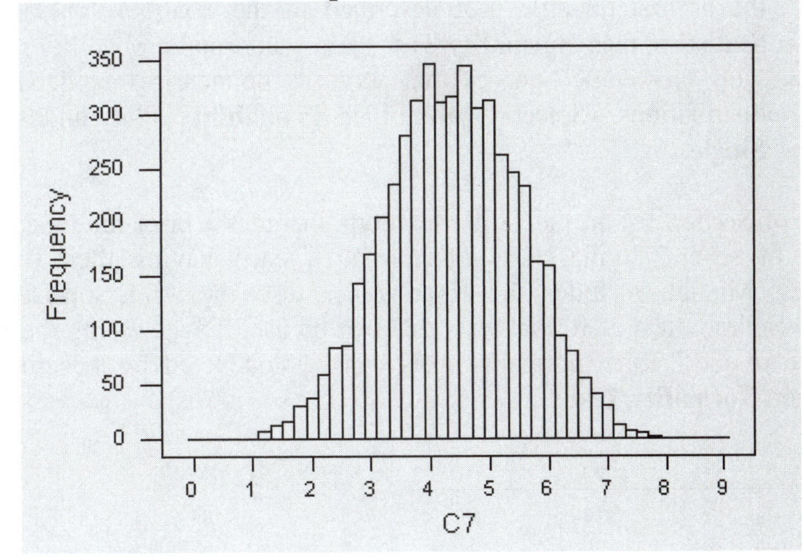

6-4 Assessing Normality

Textbooks in the Triola statistics series discuss criteria for determining whether sample data appear to come from a population having a normal distribution. (This section of Assessing Normality is not included in *Essentials of Statistics*.) These criteria are listed:

1. **Histogram:** Construct a Histogram. Reject normality if the histogram departs dramatically from a bell shape. Minitab can generate a histogram. Section 2-2 of this manual/workbook gives the Minitab procedure for generating a histogram.

2. **Outliers:** Identify outliers. Reject normality if there is more than one outlier present. (Just one outlier could be an error or the result of chance variation, but be careful, because even a single outlier can have a dramatic effect on results.) Using Minitab, we can sort the data and easily identify any values that are far away from the majority of all other values. Minitab-generated boxplots can also be used to identify potential outliers. Section 3-3 of this manual/workbook provides the Minitab procedure for generating a "modified boxplot." Modified boxplots are described in Part 2 of Section 3-4 in the textbook. In a Minitab-generated boxplot, points identified with asterisks are potential outliers.

3. **Normal Quantile Plot:** If the histogram is basically symmetric and there is at most one outlier, construct a *normal quantile plot.* Examine the normal quantile plot and reject normality if the points do not lie close to a straight line, or if the points exhibit some systematic pattern that is not a straight-line pattern. Minitab can generate a *normal probability plot*, which can be interpreted the same way as the normal quantile plot described in the textbook. (Select **Stat**, then **Basic Statistics**, then **Normality Test**.) Also, a normal probability plot can be generated with "percentile" curves that serve as boundaries for departures from normal distributions. (Select **Graph**, then **Probability Plot**, and select the option of **Single**.)

Also, Part 2 of Section 6-7 in the Triola textbook includes a brief reference to the Ryan-Joiner test as one of several formal tests of normality, each having their own advantages and disadvantages. Minitab includes the Ryan-Joiner test, the Anderson-Darling test, and the Kolmogorov-Smirnov test as formal tests that can be used for assessing the normality of a data set. Any one of these three normality tests can be conducted by selecting **Stat**, then **Basic Statistics**, then **Normality Test**.

Normal Probability Plot To obtain a normal probability plot, click on the main menu item of **Stat**, then select **Basic Statistics**, then **Normality Test**. You will get a dialog box like the one shown below. The entries in the window at the left are present because the worksheet CHMOVIE (child movies) was opened so that columns with numerical data are listed. (See Example 4 in Section 6-7 of the textbook.) Click on the desired column so that its label appears in the "Variable" box. Click **OK**. The resulting normal quantile plot will be as shown on the next page.

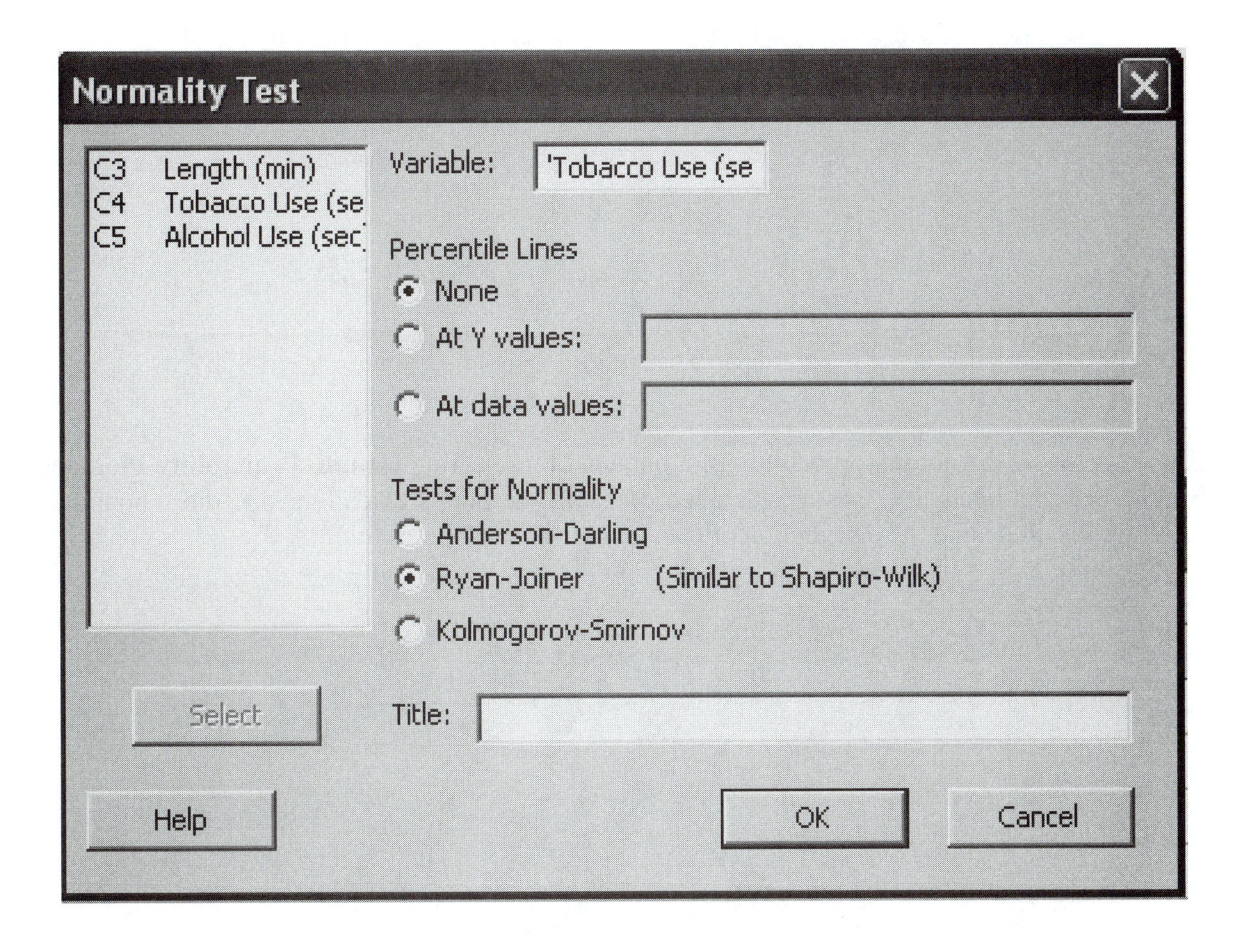

Using the normal probability plot shown below, we see that the plot does *not* yield a pattern of points that reasonably approximates a straight line, so we conclude that the times for tobacco use are not normally distributed.

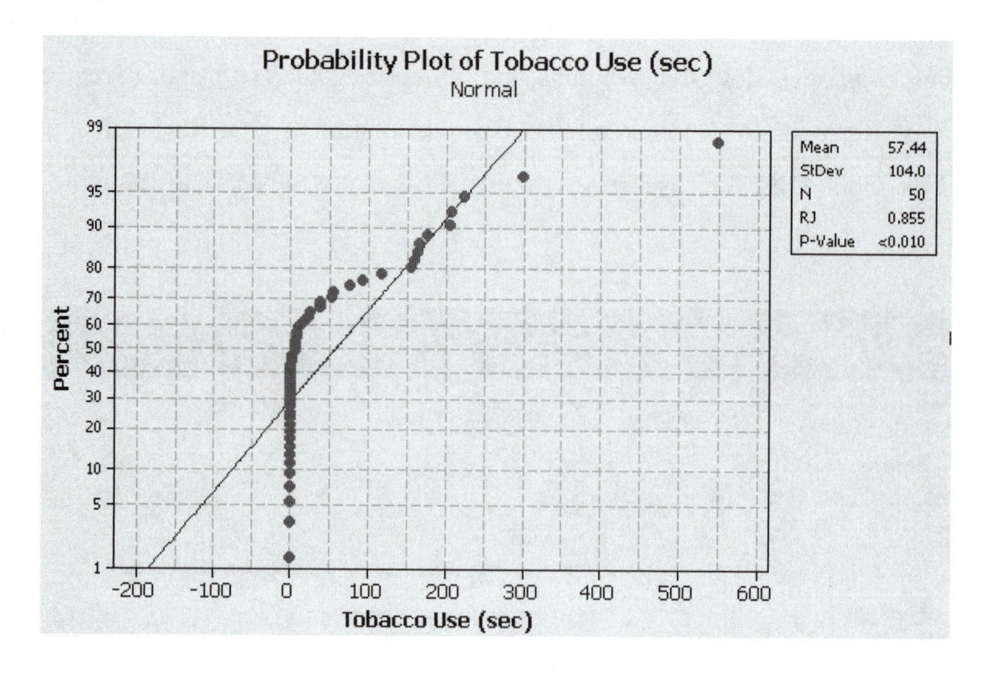

Shown below is the normal probability plot obtained by selecting **Graph**, **Probability Plot**, and **Single**. See that boundary lines are included. Because the points extend beyond those boundary lines, we conclude that the distribution of data is not a normal distribution.

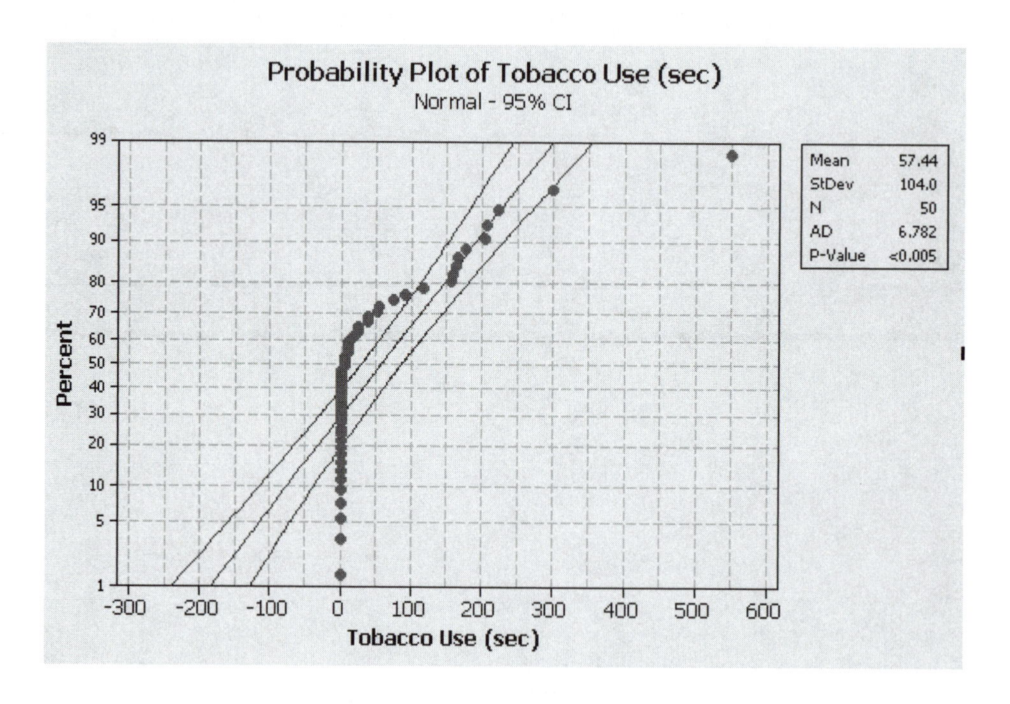

6-5 Normal Approximation to Binomial

Section 6-6 in the Triola textbook discusses the topic of using a normal distribution as an approximation to a binomial distribution. That section includes this comment:

> *Note:* Instead of using a normal distribution as an approximation to a binomial probability distribution, most practical applications of the binomial distribution can be handled with computer software or a calculator.

When working with applications involving a binomial distribution, Minitab can be used to find *exact* results, so there is no need to approximate the binomial distribution with a normal distribution. Consider Example 1 in Section 6-6 of the textbook:

> The author was mailed a survey from Viking River Cruises, and the survey included a request for an e-mail address. Assume that the survey was sent to 40,000 people and that for such surveys, the percentage of responses with an e-mail address is 3%. If the true goal of the survey was to acquire a bank of at least 1150 e-mail addresses, find the probability of getting at least 1150 responses with e-mail addresses.

Based on the above statements from Example 1 in the textbook, we have $n = 40{,}000$ and $p = 0.03$, and we want to find $P(x \geq 1150)$, which is the probability of getting at least 1150 responses with e-mail addresses. Using Minitab with the binomial distribution procedure described in Section 5-2 of this manual/workbook, we can find that the probability of 1149 or fewer responses with e-mail addresses is 0.0686665, so the *exact* probability of at least 1150 such responses is $1 - 0.0686665 = 0.9313335$. This is a more accurate result than the result of 0.9306 found by using the normal distribution as an approximation to the binomial distribution. For Minitab users, the method of approximating a binomial distribution with a normal distribution is generally obsolete.

CHAPTER 6 EXPERIMENTS: Normal Distributions

6-1. ***Finding Probabilities for a Normal Distribution*** Use Minitab's **Normal** probability distribution module to find the indicated probabilities. First select **Calc** from the main menu, then select **Probability Distributions**, then **Normal**. Probabilities can be found by using the **Cumulative Probability** option. Remember that like Table A-2 in the textbook, Minitab's probabilities correspond to cumulative areas from the left.

a. Given a population with a normal distribution, a mean of 0, and a standard deviation of 1, find the probability of a value less than 0.75.________________

b. Given a population with a normal distribution, a mean of 100, and a standard deviation of 15, find the probability of a value less than 83.________________

c. Given a population with a normal distribution, a mean of 120, and a standard deviation of 10, find the probability of a value greater than 135.______________

d. Given a population with a normal distribution, a mean of 98.6, and a standard deviation of 0.62, find the probability of a value between 97.0 and 99.0.__________

e. Given a population with a normal distribution, a mean of 150, and a standard deviation of 12, find the probability of a value between 147 and 160.__________

6-2. ***Finding Values for a Normal Distribution*** Use Minitab's **Normal** module to find the indicated vales. First select **Calc** from the main menu, then select **Probability Distributions**, then **Normal**. Values can be found by using the **Inverse cumulative probability** option. Remember that like Table A-2 in the textbook, Minitab's probabilities correspond to cumulative areas from the left.

a. Given a population with a normal distribution, a mean of 0, and a standard deviation of 1, what value has an area of 0.4 to its left?__________________

b. Given a population with a normal distribution, a mean of 100, and a standard deviation of 15, what value has an area of 0.3 to its right?________________

c. Given a population with a normal distribution, a mean of 85, and a standard deviation of 8, what value has an area of 0.95 to its left?________________

d. Given a population with a normal distribution, a mean of 45.5, and a standard deviation of 3.5, what value has an area of 0.88 to its right?________________

e. Given a population with a normal distribution, a mean of 94, and a standard deviation of 6, what value has an area of 0.01 to its left?________________

6-3. ***Central Limit Theorem*** In this experiment we will illustrate the central limit theorem. See the example in Section 6-3 of this manual/workbook.

a. Use Minitab to create columns C1, C2, C3, C4, and C5 so that each column contains 250 values that are randomly generated integers between 0 and 9.

b. Use **Data/Stack** to stack all of the 1250 values in column C6. Obtain a printed copy of the histogram for the sample data in column C6. Also find the mean and standard deviation and enter those values here.

Mean:__________ Standard deviation:__________

c. Now create a column C7 consisting of the 250 sample means. The first entry of column C7 is the mean of the first entries in columns C1 through C5, the second entry of C7 is the mean of the second entries in columns C1 through C5, and so on. (Click on **Calc**, select **Row Statistics**, select **Mean**, and enter C1-C5 for the input variables. Enter C7 as the column for storing the results. Obtain a printed copy of the histogram for the sample data in column C7. Also find the mean and standard deviation and enter those values here.

Mean:__________ Standard deviation:__________

d. How do these results illustrate the central limit theorem?

____

____

6-4. ***Central Limit Theorem*** In this experiment we will illustrate the central limit theorem. See the example in Section 6-3 of this manual/workbook.

a. Use Minitab to create columns C1 through C20 in such a way that each column contains 5000 values that are randomly generated from Minitab's **Uniform** distribution with a minimum of 0 and a maximum of 5.

b. Stack all of the sample values in column C21. [See part (b) of the preceding exercise.] Obtain a printed copy of the histogram for the sample data in column C21. Also find the mean and standard deviation and enter those values here.

Mean:__________ Standard deviation:__________

c. Now create a column C22 consisting of the 5000 sample means. The first entry of column C22 is the mean of the first entries in columns C1 through C20, the second entry of C22 is the mean of the second entries in columns C1 through C20, and so on. [See part (c) of the preceding exercise.]
Obtain a printed copy of the histogram for the sample data in column C22. Also find the mean and standard deviation and enter those values below.
Mean:__________ Standard deviation:__________

d. How do these results illustrate the central limit theorem?

____

____

6-5. ***Identifying Significance*** People generally believe that the mean body temperature is 98.6°F. The body temperature data set in Appendix B of the textbook includes a sample of 106 body temperatures with these properties: The distribution is approximately normal, the sample mean is 98.20°F, and the standard deviation is 0.62°F. We want to determine whether these sample results differ from 98.6°F by a *significant* amount. One way to make that determination is to study the behavior of samples drawn from a population with a mean of 98.6 (and a standard deviation of 0.62°F and a normal distribution).

a. Use Minitab to generate 106 values from a normally distributed population with a mean of 98.6 and a standard deviation of 0.62. Use **Stat/Basic Statistics/ Display Descriptive Statistics** to find the mean of the generated sample. Record that mean here:______________

b. Repeat part a nine more times and record the 10 sample means here:

____

c. By examining the 10 sample means in part b, we can get a sense for how much sample means vary for a normally distributed population with a mean of 98.6 and a standard deviation of 0.62. After examining those 10 sample means, what do you conclude about the likelihood of getting a sample mean of 98.20? Is 98.20 a sample mean that could easily occur by chance, or is it significantly different from the likely sample means that we expect from a population with a mean of 98.6?

____

____

d. Given that researchers did obtain a sample of 106 temperatures with a mean of 98.20°F, what does their result suggest about the common belief that the population mean is 98.6°F?

____

____

6-6. ***Identifying Significance*** This experiment involves one of the data sets in Appendix B of the textbook: "Weights and Volumes of Cola." The Minitab worksheet is COLA.

a. Open the worksheet COLA and find the mean and standard deviation of the sample consisting of the volumes of cola in cans of regular Coke. Enter the results here. Sample mean:____________ Standard deviation:____________

b. Generate 10 different samples, where each sample has 36 values randomly selected from a normally distributed population with a mean of 12 oz and a standard deviation of 0.115 oz (based on the claimed volume printed on the cans and the data in Appendix B). For each sample, record the sample mean and enter it here.

____

c. By examining the 10 sample means in part b, we can get a sense for how much sample means vary for a normally distributed population with a mean of 12 and a standard deviation of 0.115. After examining those 10 sample means, what do you conclude about the likelihood of getting a sample mean like the one found for the sample volumes in Appendix B? Is the mean for the sample a value that could easily occur by chance, or is it significantly different from the likely sample means that we expect from a population with a mean of 12?

____

d. Consider the sample mean found from the volumes of regular Coke listed in Appendix B from the textbook. Does it suggest that the population mean of 12 oz (as printed on the label) is not correct?

____

Assessing ***Normality.*** *In the following exercises, refer to the indicated data set and use Minitab to determine whether the data have a normal distribution. Give a reason explaining your choice.*

6-7. ***Space Shuttle Flights*** The lengths (in hours) of flights of NASA's Space Transport System (Shuttle) as listed in Data Set 10 in Appendix B. The Minitab worksheet is NASA.

6-8. ***Astronaut Flights*** The numbers of flights by NASA astronauts, as listed in Data Set 10 in Appendix B. The Minitab worksheet is NASA.

6-9. ***Heating Degree Days*** The values of heating degree days, as listed in Data Set 12 in Appendix B. The Minitab worksheet is ELECTRIC.

6-10. ***Generator Voltage*** The measured voltage levels from a generator, as listed in Data Set 13 in Appendix B. The Minitab worksheet is VOLTAGE.

7

Confidence Intervals and Sample Sizes

7-1 Confidence Intervals for Estimating p

7-2 Confidence Intervals for Estimating μ

7-3 Confidence Intervals for Estimating σ

7-4 Determining Sample Sizes

7-1 Confidence Intervals for Estimating *p*

Section 7-2 of the Triola textbook presents methods for using a sample proportion to estimate a population proportion, and a confidence interval is a common and important tool used for that purpose. A **confidence interval** (or **interval estimate**) is a range (or an interval) of values used to estimate the true value of a population parameter.

Important note about the Minitab methods: Section 7-2 in the textbook gives a method for constructing a confidence interval estimate of a population proportion p. The textbook procedure is based on the use of a normal distribution as an approximation to a binomial distribution. However, *Minitab uses a different method that is much more complex, but Minitab's method should provide better results.* Minitab does provide an option for using the normal approximation method.

- **Minitab's default method for generating a confidence interval estimate of a population proportion p is different from the normal approximation described in Section 7-2 of the textbook.**

- **Minitab's method for generating a confidence interval for p includes an option for using the same normal distribution approximation method described in Section 7-2 of the textbook.**

Finding x: Minitab can use the sample size n and the number of successes x to construct a confidence interval estimate of a population proportion p. In some cases, the values of x and n are both known, but in other cases the given information may consist of n and a sample percentage. For example, suppose we know that among 1007 people surveyed, 85% said that they know what Twitter is. Based on that information, we know that $n = 1007$ and $\hat{p} = 0.85$. Finding the value of x is quite simple: 85% of 1007 is $0.85 \times 1007 = 856$ (rounded to a whole number). (Because $\hat{p} = x/n$, it follows that $x = \hat{p}n$, so the number of successes can be found by multiplying the sample proportion p and the sample size n.)

To find the number of successes x from the sample proportion and sample size: Calculate $x = \hat{p} \cdot n$ and round the result to the nearest whole number.

After having determined the value of the sample size n and the number of successes x, we can proceed to use Minitab as follows.

Minitab Procedure for Obtaining Confidence Intervals for *p*

1. Select **Stat** from the main menu.

2. Select **Basic Statisticss**.

3. Select **1 Proportion**.

4. You will get a dialog box like the one shown below. The entries correspond to $n = 1007$ and $x = 856$.

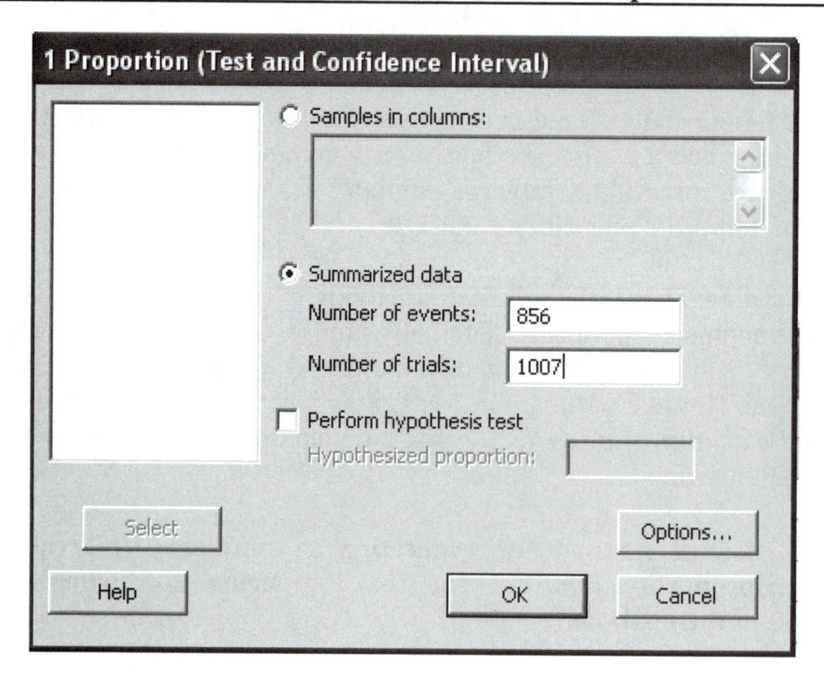

5. Click on **Options** to change the confidence level from 95% to any other value. The **Options** box can also be used to check the box for "Test and interval based on normal distribution." You could click on that box to use the same procedure given in the textbook, as shown in the following display.

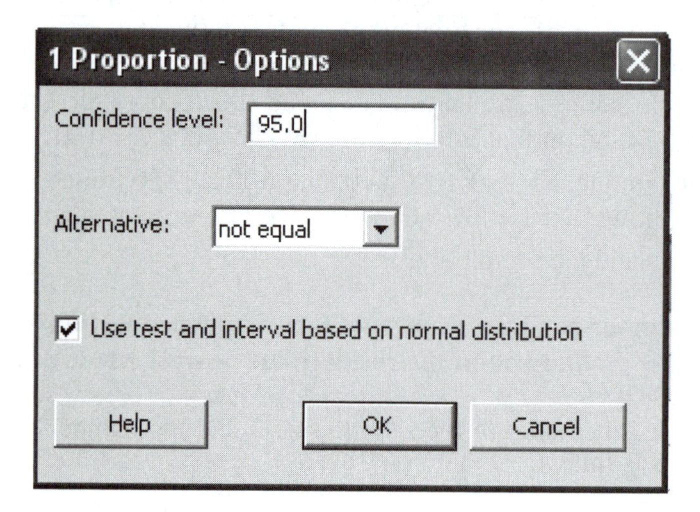

After clicking **OK**, the Minitab results are as shown below.

Test and CI for One Proportion

```
Sample     X     N  Sample p         95% CI
1        856  1007  0.850050  (0.827999, 0.872101)

Using the normal approximation.
```

Based on the above Minitab display, we can express the 95% confidence interval estimate of p as $0.827999 < p < 0.872101$. After rounding, the confidence interval becomes $0.828 < p < 0.872$. If you do not select the option of using the normal distribution approximation method, Minitab will provide the confidence interval of $0.826 < p < 0.872$. In this case, the result is very close to the result obtained using the normal distribution approximation method described in the textbook.

7-2 Confidence Intervals for Estimating μ

Section 7-3 in the textbook introduces a method for using a sample mean $\bar{x}$ to estimate the value of a population mean μ. Example 2 in Section 7-3 provides the following speeds (mi/h) measured from southbound traffic on I-280 near Cupertino, California (based on data from SigAlert).

62 61 61 57 61 54 59 58 59 69 60 67

Part 1 of Section 7-3 in the textbook focuses on the case in which the population standard deviation σ is not known, as is usually the case. Part 2 of Section 7-3 in the textbook discusses the case in which σ is known. See the textbook for the requirements for each method. The requirements are summarized in Table 7-1 in the textbook. Here is the procedure for using Minitab to construct confidence interval estimates of a population mean μ.

Minitab Procedure for Obtaining Confidence Intervals for μ

1. Select **Stat** from the main menu.
2. Select **Basic Statistics** from the subdirectory.
3. Select **1-Sample t** (or select **1-Sample** z if σ is known.)
4. You will now see a dialog box, such as the one shown below. For the sample data, you can use either the original list of sample values by selecting **Samples in** columns, or you can use the summary statistics by selecting **Summarized** data. (If you select 1-Sample z in Step 3, there will also be a box for entering "Sigma," the population standard deviation σ.) Enter the label of the column containing the list of data or enter the summary statistics (sample size n, sample mean $\bar{x}$, sample standard deviation s). In the following screen display, the column named “Speeds” contains the 12 speeds listed above.

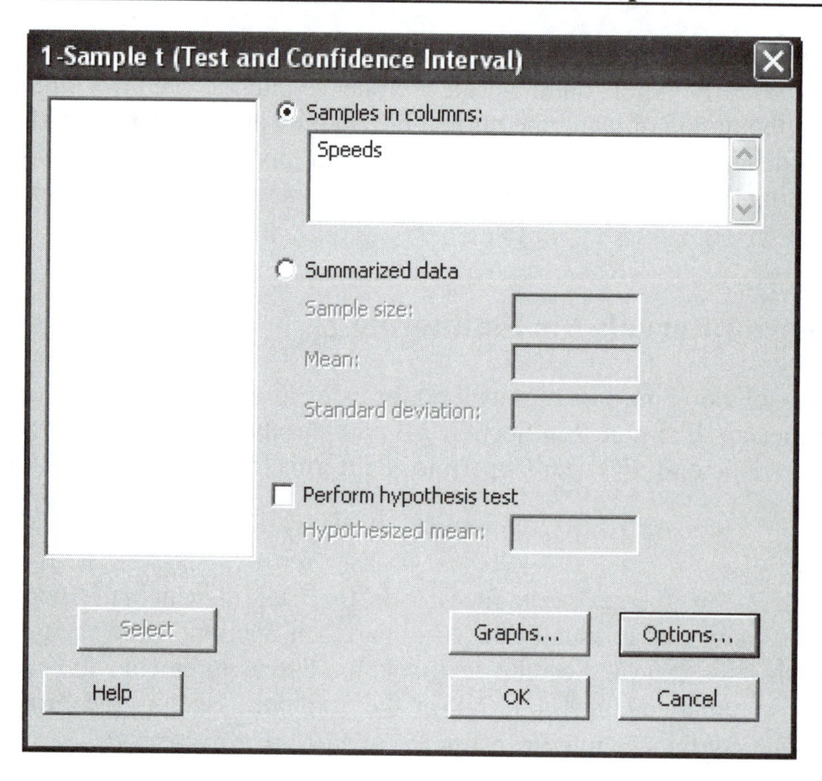

5. Click on the **Options** box to enter the confidence level. (The Minitab default is a confidence level of 95%.) Click **OK** when done.

If you click **OK** on the above screen, you get the following Minitab display.

One-Sample T: Speeds

Variable	N	Mean	StDev	SE Mean	95% CI
Speeds	12	60.67	4.08	1.18	(58.08, 63.26)

Based on the above Minitab display, the 95% confidence interval is $58.1 < \mu < 63.3$ (rounded). The solution in the textbook includes this statement: We are 95% confident that the limits of 58.1 mi/h and 63.3 mi/h actually do contain the value of the population mean μ.

7-3 Confidence Intervals for Estimating σ

Here is the Minitab procedure for constructing confidence interval estimates of σ or σ^2.

Minitab Procedure for Confidence Intervals for Estimating σ

1. Enter the original sample values in a Minitab column, or obtain the values of the sample size n and the sample standard deviation s (or the sample variance s^2)..
2. Select **Stat** from the main menu.
3. Select **Basic Statistics** from the subdirectory.
4. Select **1 Variance**.
5. You will get a dialog box. Click on **Data** and select one of these three options:
   Samples in columns (Sample data are listed in a column)
   Sample standard deviation (The value of s is known.)
   Sample variance (The value of s^2 is known.)
6. Either enter the label of the column containing the sample data or enter the required sample statistics, then click **OK**.

Example 2 from Section 7-4 in the textbook states that a sample of size $n = 22$ has standard deviation $s = 14.3$, and we want to construct a 95% confidence interval estimate of σ. Because we know the values of $n = 22$ and $s = 14.3$, select **Sample standard deviation** in Step 5. You will now get the dialog box given below, so enter the values of the sample statistics as shown .

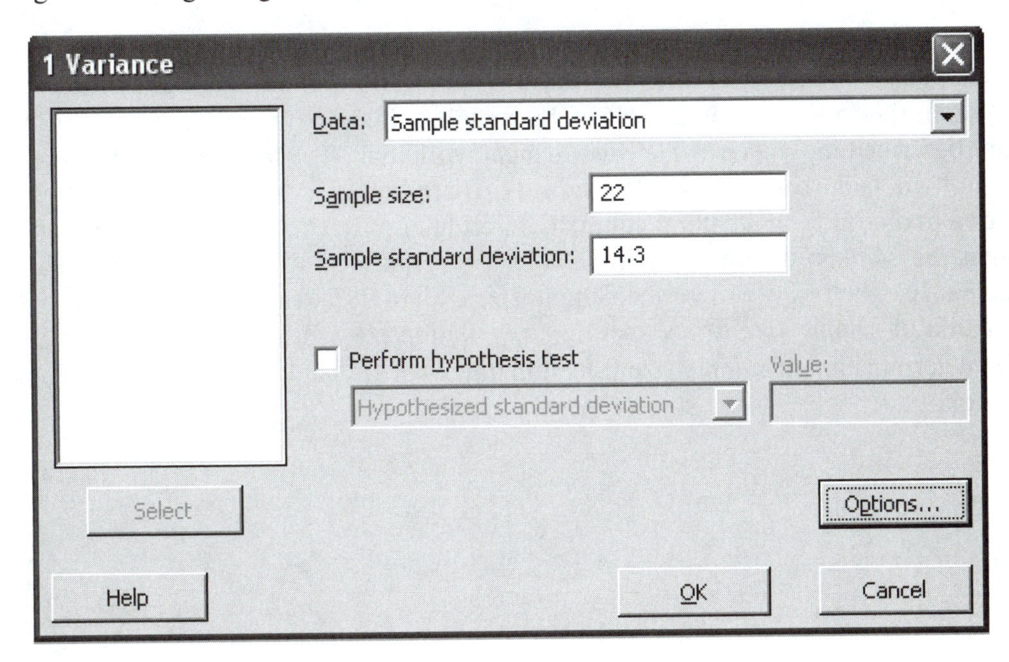

After clicking on **OK**, the following results will be displayed.

```
95% Confidence Intervals

                   CI for        CI for
Method             StDev         Variance
Chi-Square    (11.0, 20.4)    (121, 418)
```

Based on this Minitab result, we have this 95% confidence interval estimate of the population standard deviation: $11.0 < \sigma < 20.4$.

7-4 Determining Sample Sizes

The textbook discuss procedures for determining sample sizes necessary to estimate a population proportion p, or mean μ, or standard deviation σ. Minitab does not have the ability to calculate sample sizes directly, but determinations of sample size use simple formulas, so it is easy to use a calculator with the procedures described in the textbook. Minitab could be used instead of a calculator. Consider this formula for determining the sample size required to estimate a population mean:

$$n = \left[\frac{z_{\alpha/2}\sigma}{E}\right]^2$$

Suppose that we want to find the sample size needed to estimate the mean IQ score of statistics professors, and we want 95% confidence that the error in the sample mean is no more than 2 IQ points. We further assume that $\sigma = 15$. The required value of z can be found by using the procedure described in Section 6-1 of this manual/workbook. To find the z score corresponding to 95% in two tails, select **Calc, Probability Distributions, Normal,** then select **Inverse cumulative probability** and enter a cumulative left area of 0.975 to get the z score of 1.95996. Then enter the Minitab command LET C2=(1.95996*15/2)**2 to complete the calculation for the sample size. The result will be the sample size of 216.082, which we round up to 217. The determination of sample size for estimating a population proportion p can be handled the same way. The determination of sample size for estimating σ or σ^2 can be done by using Table 7-2 in the textbook.

CHAPTER 7 EXPERIMENTS: Confidence Intervals and Sample Sizes

In Experiments 7–1 through 7–4, use Minitab with the sample data and confidence level to construct the confidence interval estimate of the population proportion p.

7-1. From a KRC Research poll in which respondents were asked if they felt vulnerable to identify theft: $n = 1002$, $x = 531$ who said "yes". Use a 95% confidence level.

____

7-2. From a 3M Privacy Filters poll in which respondents were asked to identify their favorite seat when they fly: $n = 806$, $x = 492$ who chose the window seat. Use a 99% confidence level.

____

7-3. From a Prince Market Research poll in which respondents were asked if the acted to annoy a bad driver: $n = 2518$, $x = 1083$ who said that they honked. Use a 90% confidence level.

____

7-4. From an Angus Reid Public Opinion poll in which respondents were asked if they felt that U. S. nuclear weapons made them feel safer: $n = 1005$, $x = 543$ who said "yes". Use a confidence level of 80%.

____

In Experiments 7–5 through 7–10, use Minitab with the sample data and confidence level to construct the confidence interval estimate of the population mean μ. *Be careful to correctly choose between the t and normal distributions.*

7–5. $n = 50$, $\bar{x} = 75$, $s = 12$, 95% confidence _______________

7–6. $n = 87$, $\bar{x} = 62.3$, $s = 6.6$, 95% confidence _______________

7–7. $n = 32$, $\bar{x} = 8.92$, $s = 1.33$, 99% confidence _______________

7–8. $n = 106$, $\bar{x} = 98.2$, $s = 0.62$, 99% confidence _______________

7–9. $n = 81$, $\bar{x} = 12.8$, $s = 2.7$, 95% confidence _______________

7–10. $n = 212$, $\bar{x} = 1008.1$, $s = 47.3$, 95% confidence _______________

In Experiments 7–11 through 7–14, use Minitab to construct the confidence interval estimate of the population mean μ. In each case, the population standard deviation σ is known. Be careful to correctly choose between the t and normal distributions.

7–11. $n = 212$, $\bar{x} = 625$, $\sigma = 38$, 95% confidence ____________________

7–12. $n = 53$, $\bar{x} = 57$, $\sigma = 7$, 95% confidence ____________________

7–13. $n = 71$, $\bar{x} = 82.4$, $\sigma = 11.2$, 95% confidence ____________________

7–14. $n = 150$, $\bar{x} = 9.9$, $\sigma = 2.7$, 95% confidence ____________________

7–15. ***Mendelian Genetics*** When Mendel conducted his famous genetics experiments with peas, one sample of offspring consisted of 428 green peas and 152 yellow peas.

a. Use Minitab to find the following confidence interval estimates of the percentage of yellow peas.
99.5% confidence interval: ____________________
99% confidence interval: ____________________
98% confidence interval: ____________________
95% confidence interval: ____________________
90% confidence interval: ____________________

b. After examining the pattern of the above confidence intervals, complete the following statement. "As the degree of confidence decreases, the confidence interval limits
____."

c. In your own words, explain why the preceding completed statement makes sense. That is, why should the confidence intervals behave as you have described?

____

____

7-16. ***Gender Selection*** The Genetics and IVF Institute conducted a clinical trial of the XSORT method designed to increase the probability of conceiving a girl. As of this writing, 945 babies were born to parents using the XSORT method, and 879 of them were girls.

a. What is the best point estimate of the population proportion of girls born to parents using the XSORT method? ____________________

b. Use the sample data to construct a 95% confidence interval estimate of the percentage of girls born to parents using the XSORT method.

____

(*continued*)

c. Based on the results, does the XSORT method appear to be effective? Why or why not?

______

______

______

7-17. ***Gender Selection*** The Genetics and IVF Institute conducted a clinical trial of the YSORT method designed to increase the probability of conceiving a boy. As of this writing, 291 babies were born to parents using the YSORT method, and 239 of them were boys.

a. What is the best point estimate of the population proportion of boys born to parents using the YSORT method? ______

b. Use the sample data to construct a 99% confidence interval estimate of the percentage of boys born to parents using the YSORT method.

______

c. Based on the results, does the YSORT method appear to be effective? Why or why not?

______

______

7-18. ***Touch Therapy*** When she was nine years of age, Emily Rosa did a science fair experiment in which she tested professional touch therapists to see if they could sense her energy field. She flipped a coin to select either her right hand or her left hand, then she asked the therapists to identify the selected hand by placing their hand just under Emily's hand without seeing it and without touching it. Among 280 trials, the touch therapists were correct 123 times (based on data in "A Close Look at Therapeutic Touch," *Journal of the American Medical Association*, Vol. 279, No. 13).

a. Given that Emily used a coin toss to select either her right hand or her left hand, what proportion of correct responses would be expected if the touch therapists made random guesses? ______

b. Using Emily's sample results, construct a 99% confidence interval estimate of the proportion of correct responses made by touch therapists.

______

c. What do the results suggest about the ability of touch therapists to select the correct hand by sensing an energy field?

______

In Exercises 7-19 through 7-22, use the data sets from Appendix B in the textbook.

7-19. ***Nicotine in Cigarettes*** Refer to Data Set 10 in Appendix B of the textbook and assume that the samples are simple random samples obtained from normally distributed populations. The Minitab worksheet is CIGARET.

a. Construct a 95% confidence interval estimate of the mean amount of nicotine in cigarettes that are king size, non-filtered, non-menthol, and non-light.

____

b. Construct a 95% confidence interval estimate of the mean amount of nicotine in cigarettes that are 100 mm, filtered, non-menthol, and non-light.

____

c. Compare the results. Do filters on cigarettes appear to be effective?

____

____

7-20. ***Pulse Rates*** A physician wants to develop criteria for determining whether a patient's pulse rate is atypical, and she wants to determine whether there are significant differences between males and females. Use the sample pulse rates in Data Set 1 from Appendix B in the textbook. The Minitab worksheets are MBODY and FBODY.

a. Construct a 95% confidence interval estimate of the mean pulse rate for males.

____

b. Construct a 95% confidence interval estimate of the mean pulse rate for females.

____

c. Compare the preceding results. Can we conclude that the population means for males and females are different? Why or why not?

____

____

____

7-21. ***Weights of Coins*** Refer to the weights of pre-1964 quarters and the weights of post-1964 quarters listed in Data Set 21 from Appendix B of the textbook. The Minitab worksheet is COINS.

a. Construct a 95% confidence interval estimate of the mean weight of pre-1964 quarters.

____

b. Construct a 95% confidence interval estimate of the mean weight of post-1964 quarters.

____

c. Compare the preceding results. Can we conclude that the population means for pre-1964 quarters and post-1964 quarters are different? Why or why not?

____

____

7-22. ***Second Hand Smoke*** Refer to the measured cotinine levels of smokers, nonsmokers exposed to cigarette smoke, and nonsmokers not exposed to cigarette smoke. The cotinine measurements are listed in Data Set 9 from Appendix B of the textbook. The Minitab worksheet is COTININE.

a. Construct a 95% confidence interval estimate of the mean cotinine level of smokers.

____

b. Construct a 95% confidence interval estimate of the mean cotinine level of nonsmokers exposed to cigarette smoke. (The column label is ETS.)

____

c. Construct a 95% confidence interval estimate of the mean cotinine level of nonsmokers not exposed to cigarette smoke. (The column label is NOETS.)

____

d. Compare the preceding results. What do you conclude?

____

____

____

7-23. ***Simulated Data*** Minitab is designed to generate random numbers from a variety of different sampling distributions. In this experiment we will generate 500 IQ scores, then we will construct a confidence interval based on the sample results. IQ scores have a normal distribution with a mean of 100 and a standard deviation of 15. First generate the 500 sample values as follows.

1. Click on **Calc,** then select **Random Data**, then **Normal**.
2. In the dialog box, enter a sample size of 500, a mean of 100, and a a standard deviation of 15. Specify column C1 as the location for storing the results. Click **OK**.
3. Use **Stat/Basic Statistics/Display Descriptive Statistics** to find these statistics:

$n =$ __________ $\bar{x} =$ __________ $s =$ __________

Using the generated values, construct a 95% confidence interval estimate of the population mean of all IQ scores. Enter the 95% confidence interval here.

____

Because of the way that the sample data were generated, we *know* that the population mean is 100. Do the confidence interval limits contain the true mean IQ score of 100?___

If this experiment were to be repeated many times, how often would we expect the confidence interval limits to contain the true population mean value of 100? Explain how you arrived at your answer.

____

____

7-24. ***Simulated Data*** Follow the same steps listed in Experiment 7-23 to randomly generate 500 IQ scores from a population having a normal distribution, a mean of 100, and a standard deviation of 15. Record the sample statistics here.

$n =$ __________ $\bar{x} =$ __________ $s =$ __________

Confidence intervals are typically constructed with confidence levels around 90%, 95%, or 99%. Instead of constructing such a typical confidence interval, use the generated values to construct a 50% confidence interval. Enter the result below.

____

Does the above confidence interval have limits that actually do contain the true population mean, which we know is 100? __________

(continued)

Repeat the above procedure 9 more times and list the resulting 50% confidence intervals here.

______________________ ______________________

______________________ ______________________

______________________ ______________________

______________________ ______________________

____

Among the total of the 10 confidence intervals constructed, how many of them actually do contain the true population mean of 100? Is this result consistent with the fact that the level of confidence used is 50%? Explain.

____

____

7-25. ***Combining Data Sets*** Refer to the M&M Data Set 20 in Appendix B of the textbook and use the entire sample of 100 plain M&M candies to construct a 95% confidence interval for the mean weight of all M&Ms. The Minitab worksheet is M&M. (*Hint*: It is not necessary to manually enter the 100 weights, because they are already stored in separate Minitab columns in the worksheet M&M. (*Hint:* Select **Data,** then use **Stack** to combine the different M&M data sets into one big data set.) Now construct a 95% confidence interval estimate of the population mean of all IQ scores. Enter the 95% confidence interval here.

____

7-26. ***Combining Data Sets*** Refer to the word counts in Data Set 17 in Appendix B of the textbook. The Minitab worksheet is WORDS. Combine all of the word counts from males in one column, and combine all of the word counts from females in another column. (*Hint:* Select **Data**, then select **Stack.**)

a. Find the 95% confidence interval estimate of the mean word count for men.

____

b. Find the 95% confidence interval estimate of the mean word count for women.

____

c. Based on the results, does it appear the women talk more than men? Why or why not?

____

____

In Exercises 7-27 through 7-30, use the data sets from Appendix B in the textbook to construct confidence interval estimates of the population standard deviations.

7-27. ***Nicotine in Cigarettes*** Refer to Data Set 10 in Appendix B of the textbook and assume that the samples are simple random samples obtained from normally distributed populations. The Minitab worksheet is CIGARET.

a. Construct a 95% confidence interval estimate of the standard deviation of amounts of nicotine in cigarettes that are king size, non-filtered, non-menthol, and non-light. ______________________________

b. Construct a 95% confidence interval estimate of the standard deviation of amounts of nicotine in cigarettes that are 100 mm, filtered, non-menthol, and non-light. ______________________________

c. Compare the results. Do the amounts of variation appear to be different?

____

____

7-28. ***Pulse Rates*** A physician wants to develop criteria for determining whether a patient's pulse rate is atypical, and she wants to determine whether there are significant differences between males and females. Use the sample pulse rates in Data Set 1 from Appendix B in the textbook. The Minitab worksheets are MBODY and FBODY.

a. Construct a 95% confidence interval estimate of the standard deviation of pulse rates for males.

____

b. Construct a 95% confidence interval estimate of the standard deviation of pulse rates for females.

____

c. Compare the preceding results. Do the amounts of variation appear to be different?

____

____

____

____

7-29. ***Weights of Coins*** Refer to the weights of pre-1964 quarters and the weights of post-1964 quarters listed in Data Set 21 from Appendix B of the textbook. The Minitab worksheet is COINS.

a. Construct a 95% confidence interval estimate of the standard deviation of the weights of pre-1964 quarters.

____

b. Construct a 95% confidence interval estimate of the standard deviation of the weights of post-1964 quarters.

____

c. Compare the preceding results. Do the amounts of variation appear to be different?

____

____

7-30. ***Second Hand Smoke*** Refer to the measured cotinine levels of smokers, nonsmokers exposed to cigarette smoke, and nonsmokers not exposed to cigarette smoke. The cotinine measurements are listed in Data Set 9 from Appendix B of the textbook. The Minitab worksheet is COTININE.

a. Construct a 95% confidence interval estimate of the standard deviation of cotinine levels of smokers.

____

b. Construct a 95% confidence interval estimate of the standard deviation of cotinine levels of nonsmokers exposed to cigarette smoke. (The column label is ETS.)

____

c. Construct a 95% confidence interval estimate of the standard deviation of cotinine levels of nonsmokers not exposed to cigarette smoke. (The column label is NOETS.)

____

d. Compare the preceding results. What do you conclude?

____

____

____

8

Hypothesis Testing

8–1 Minitab's Assistant Feature

Minitab introduced its new *Assistant* feature in Release 16. If you click on the main menu item of **Assistant,** you get a submenu with options including these:

Graphical Analysis
Hypothesis Tests
Regression
Control Charts

If you select the submenu item of **Hypothesis Tests,** you get the following display.

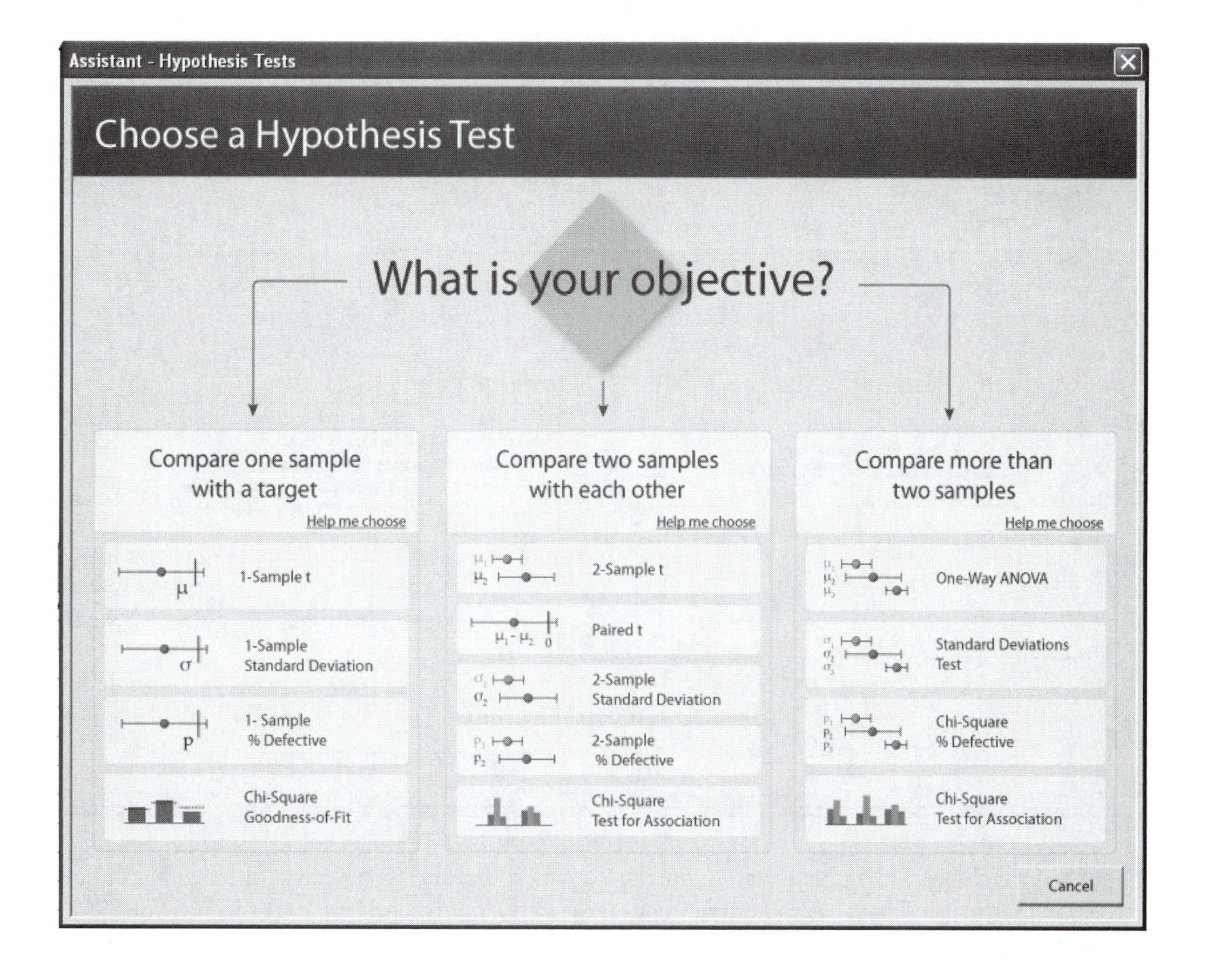

You can see that this display is helpful in guiding you to choose the correct method. For the purposes of Chapter 8 in the textbook, we would like to compare one sample with a target, so the top arrows guide us to the leftmost column of choices. (You could also click on **Help me choose** to obtain further guidance.) If you want to test a claim about a population proportion, select the item labeled **1 Sample % Defective**. You will get a dialog box that asks for more information in a simpler format, and after entering the information about the sample, you will get three screens of results that provide much information that is helpful in interpreting your results. Here is an example of one of the screens:

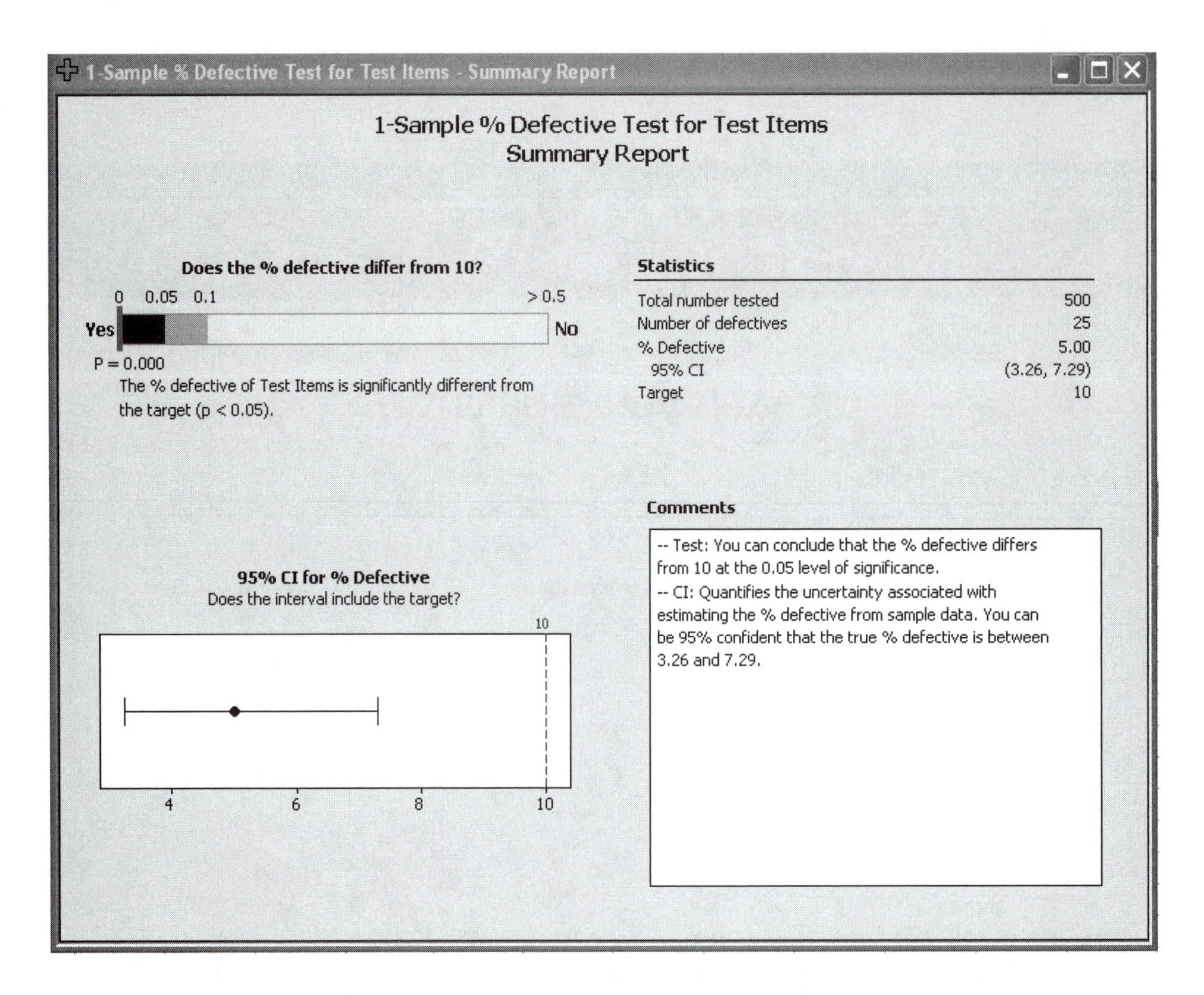

The screen includes a comment that "You can conclude that the % defective differs from 10 at the 0.05 significance level." Another comment describes a confidence interval and provides 95% confidence interval limits. In general, Minitab's *Assistant* provides helpful guidance in selecting the correct method and it provides results that are very helpful in analyzing and interpreting the results.

The following sections of this manual/workbook describe the use of Minitab's basic functions, but consider using the *Assistant* feature in place of those basic methods.

8–2 Testing Hypotheses About *p*

Important note about the Minitab methods: Section 8-3 in the textbook gives a method for testing a claim about a population proportion *p*. The textbook procedure is based on the use of a normal distribution as an approximation to a binomial distribution. However, *Minitab uses a different method that is much more complex, but Minitab's method should provide better results.* Minitab does provide an option for using the normal approximation method.

- **Minitab's default method for testing claims about a population proportion *p* is different from the normal approximation described in Section 8-3 of the textbook.**

- **Minitab's method for testing claims about *p* includes an option for using the same normal distribution approximation method described in Section 8-3 of the textbook.**

The Minitab procedure for testing claims about a population proportion follows the same basic steps described in Section 7–1 of this manual/workbook. In addition to having a claim to be tested, Minitab also requires a significance level, the sample size n, and the number of successes x.

Finding x: In Section 7-1 of this manual/workbook, we briefly discussed one particular difficulty that arises when the available information provides the sample size n and the sample proportion $\hat{p}$ instead of the number of successes x. We provided this procedure for determining the number of successes x.

To find the number of successes x from the sample proportion and sample size: Calculate $x = \hat{p}n$ and round the result to the nearest whole number.

Given a claim about a proportion and knowing n and x, we can use Minitab as follows.

Minitab Procedure for Testing Claims about *p*

1. Select **Stat** from the main menu.

2. Select **Basic Statistics** from the subdirectory.

3. Select **1 Proportion**.

4. You will get a dialog box.

   - Click on the option for summarized data.
   - In the box for "Number of trials," enter the value of the sample size n.
   - In the box for the number of events (or successes), enter the number of successes x.
   - Click on the **Options** bar and enter the confidence level. (Enter 95 for a 0.05 significance level.) Also, in the "Test proportion" box, enter the claimed value of the population proportion p. Also select the format of the alternative hypothesis.
   - Click **OK**.

For example, let's use the sample data of $n = 400$ and $x = 380$ to test the claim that $p = 0.93$. (See Example 1 in Section 8-3 of the textbook.) The dialog boxes and Minitab results are as follows.

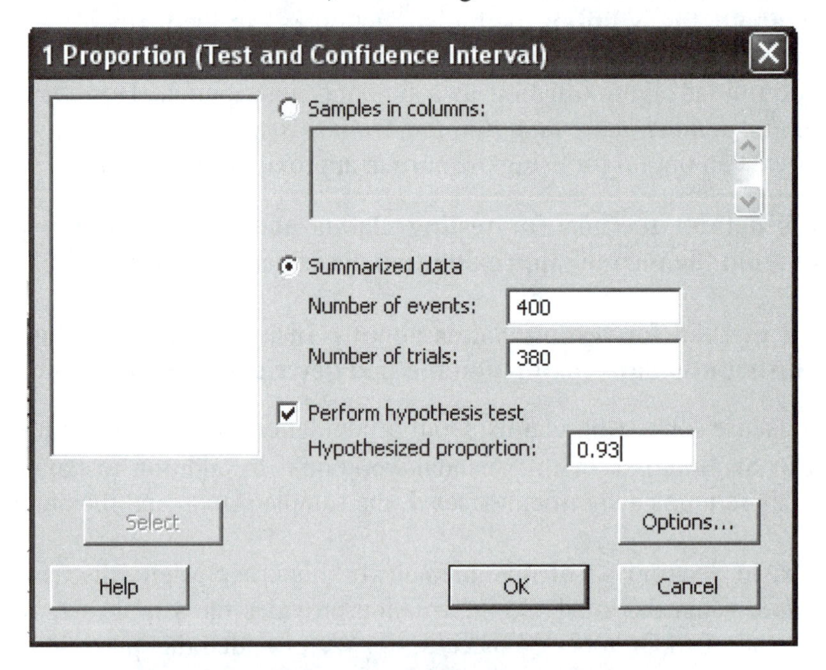

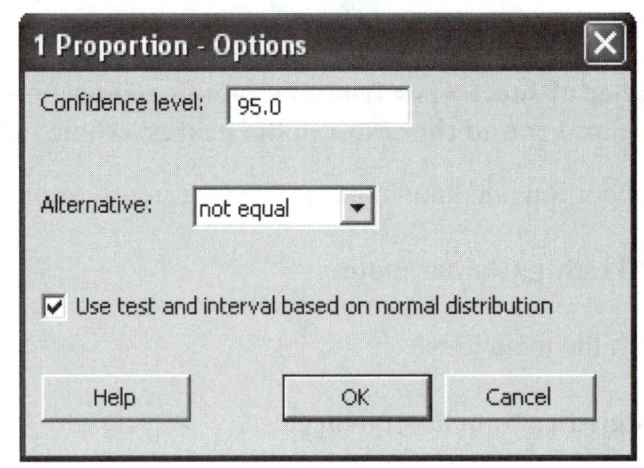

Based on the above dialog boxes, the following Minitab results are obtained.

Test and CI for One Proportion

```
Test of p = 0.93 vs p not = 0.93

Sample    X    N  Sample p          95% CI           Z-Value  P-Value
1       380  400  0.950000  (0.928642, 0.971358)      1.57    0.117

Using the normal approximation.
```

We can see from this display that the *P*-value is 0.117, which causes us to fail to reject the null hypothesis. There is not sufficient evidence to warrant rejection of the claim that $p = 0.93$. (See Example 1 in Section 8-3 of the textbook.)

Exact Method: If we had not told Minitab to conduct the test using the normal distribution approximation method (by not checking the box in the preceding screen), we would obtain the following results showing that the exact *P*-value is 0.118, and this result is better than the one obtained using the normal distribution approximation method.

Test and CI for One Proportion

```
Test of p = 0.93 vs p not = 0.93

                                                          Exact
Sample    X    N  Sample p          95% CI               P-Value
1       380  400  0.950000  (0.923833, 0.969195)           0.118
```

8–3 Testing Hypotheses About μ

Section 8-4 in the textbook describes details for methods of testing claims about a population mean μ. Part 1 of Section 8-4 focuses on the realistic case in which the population standard deviation σ is not known, and Part 2 briefly considers the somewhat artificial case in which σ is somehow known.

Minitab allows us to test claims about a population mean μ by using either the list of original sample values or the summary statistics of n, $\bar{x}$, and s.

Minitab Procedure for Testing Claims About μ

1. Select **Stat** from the main menu.

2. Select **Basic Statistics** from the subdirectory.

3. Select either **1-Sample t** (or **1-Sample z** if σ is known).

4. You will now see a dialog box, such as the one shown below. (If you select 1-Sample z in Step 4, there will also be a box for entering "Sigma," the population standard deviation σ.)

5. In the "Samples in columns" box, enter the column containing the list of sample data, or if the summary statistics are known, click **Summarized data** and proceed to enter the sample size, sample mean, and sample standard deviation.

6. Click on the box labeled "Perform hypothesis test."

7. In the "Hypothesized mean" box, enter the claimed value of the population mean.

8. Click on the **Options** button to get the options dialog box, and make the entries according to the following.

   - Enter a confidence level. (Enter 95 for a hypothesis test conducted with a 0.05 significance level.)

   - Click on the "Alternative" box to change the default alternative hypothesis. The available choices in the "Alternative" box are

     less than
     not equal
     greater than

   - Click **OK** when done.

9. Click **OK** on the main dialog box.

As an illustration, consider this, from Example 1 in Section 8-4:

> **Cell Phone Radiation** Listed below are the measured radiation emissions (in W/kg) corresponding to a sample of cell phones (based on data are from the Environmental Working Group). Use a 0.05 significance level to test the claim that cell phones have a mean radiation level that is less than 1.00 W/kg.
>
> 0.38 0.55 1.54 1.55 0.50 0.60 0.92 0.96 1.00 0.86 1.46

Because $n \leq 30$, we must verify that the sample data appear to be from a normally distributed population, and Example 1 in Section 8-4 of the textbook shows the normal quantile plot suggesting that this requirement is met. Because σ is not known, use Minitab's **1-Sample t** (instead of 1-Sample z). First enter the sample data in column C1, then proceed to conduct the t test by using the following dialog boxes that correspond to the above example. See that the first dialog box includes the entry of column C1 containing the list of sample values and the box for conducting a hypothesis test is checked, and the value of 1.00 is entered. (The claim of a mean "less than 1.00 W/kg" indicates that the null hypothesis is H_0: μ = 1.00 W/kg, so 1.00 is the hypothesized mean.) The second dialog box shows the confidence level of 95, which corresponds to a 0.05 significance level, and the second dialog box also shows the choice of "less than" for the form of the alternative hypothesis.

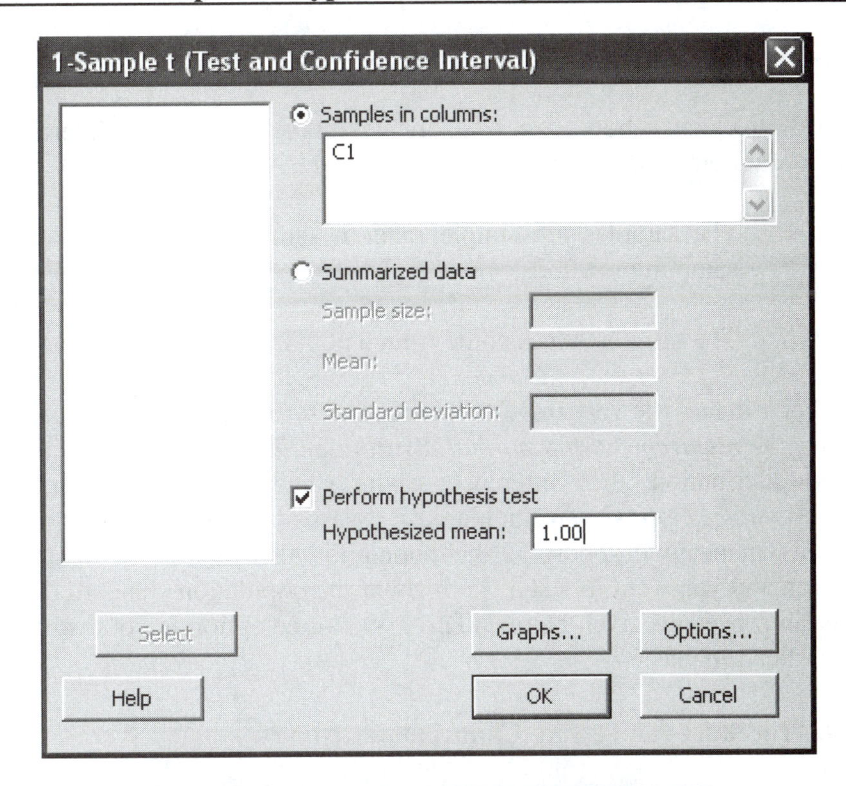

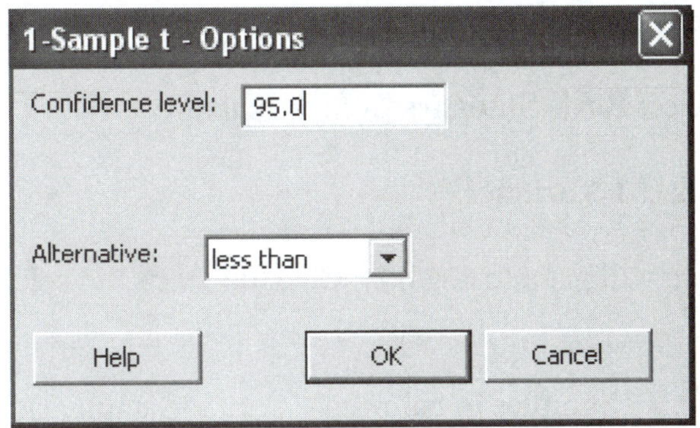

We get these Minitab results:

One-Sample T: C1

```
Test of mu = 1 vs < 1

                                          95% Upper
Variable   N   Mean  StDev  SE Mean      Bound      T      P
C1        11  0.938  0.423    0.127      1.169  -0.48  0.319
```

From the above Minitab display, we see that the test statistic is $t = -0.48$. The P-value is 0.319. Because the P-value is greater than the significance level of 0.05, we fail to reject the null hypothesis and conclude that there is not sufficient evidence to support the claim that the population mean is less than 1.00 W/kg. (See Example 1 in Section 8-4 in the textbook.)

8-4 Testing Hypotheses About σ or σ^2

When testing claims about σ or σ^2, the textbook stresses that there are the following two important requirements:

1. The samples are simple random samples. (Remember the importance of good sampling methods.)

2. The sample values come from a population with a *normal distribution.*

The textbook makes the very important point that *for tests of claims about standard deviations or variances, the requirement of a normal distribution is very strict*. If the population does not have a normal distribution, then inferences about standard deviations or variances can be very misleading. *Suggestion:* Given sample data, assess normality by using the methods from Section 6-4 of this manual/workbook. If the population distribution does appear to have a normal distribution and you want to test a claim about the population standard deviation or variance, use the Minitab procedure given below. (The **1 Variance** option is not available on Minitab Release 14 and earlier versions.)

Minitab Procedure for Testing Claims about σ or σ^2

1. Select **Stat** from the main menu.

2. Select **Basic Statistics** from the subdirectory.

3. Select **1 Variance**.

4. You will get a dialog box. Click on **Data** and select one of these three options:

Samples in columns	(Sample data are listed in a column)
Sample standard deviation	(The value of s is known.)
Sample variance	(The value of s^2 is known.)

5. Either enter the label of the column containing the sample data or enter the required sample statistics, then click **OK**.

6. Click on the box labeled "Perform hypothesis test."

7. In the "Hypothesized standard deviation" box, enter the claimed value of the population standard deviation.

8. Click on the **Options** button to get the options dialog box, and make the entries according to the following.

   - Enter a confidence level. (Enter 95 for a hypothesis test conducted with a 0.05 significance level.)
   - Click on the "Alternative" box to change the default alternative hypothesis. The available choices in the "Alternative" box are
     less than
     not equal
     greater than
   - Click **OK** when done.

9. Click **OK** on the main dialog box.

As an illustration, consider this from Example 1 in Section 8-5:

> **Supermodel Heights** Listed below are the heights (inches) for the simple random sample of supermodels. Consider the claim that supermodels have heights that have much less variation than heights of women in the general population. We will use a 0.01 significance level to test the claim that supermodels have heights with a standard deviation that is less than 2.6 in. for the population of women.
>
> 70 71 69.25 68.5 69 70 71 70 70 69.5

From the above example, we see that we want to test the claim that $\sigma < 2.6$ in., and we want to use a 0.01 significance level. See Example 1 in Section 8-5 of the textbook, where the normality of the distribution is verified with a normal quantile plot. We can proceed with the hypothesis test, and the Minitab dialog boxes are shown below, followed by the test results.

In the first dialog box, see that we chose the option of "Samples in columns" because the data were entered as a list in column C1 of Minitab. Also, the "Perform hypothesis test" box is checked, and the value of 2.6 is entered in the "Value" box. (The value of 2.6 is from the claim that $\sigma < 2.6$ in.)

In the second dialog box, the confidence level of 99 was entered (for the significance level of 0.01), and the format of "less than" was chosen for the alternative hypothesis.

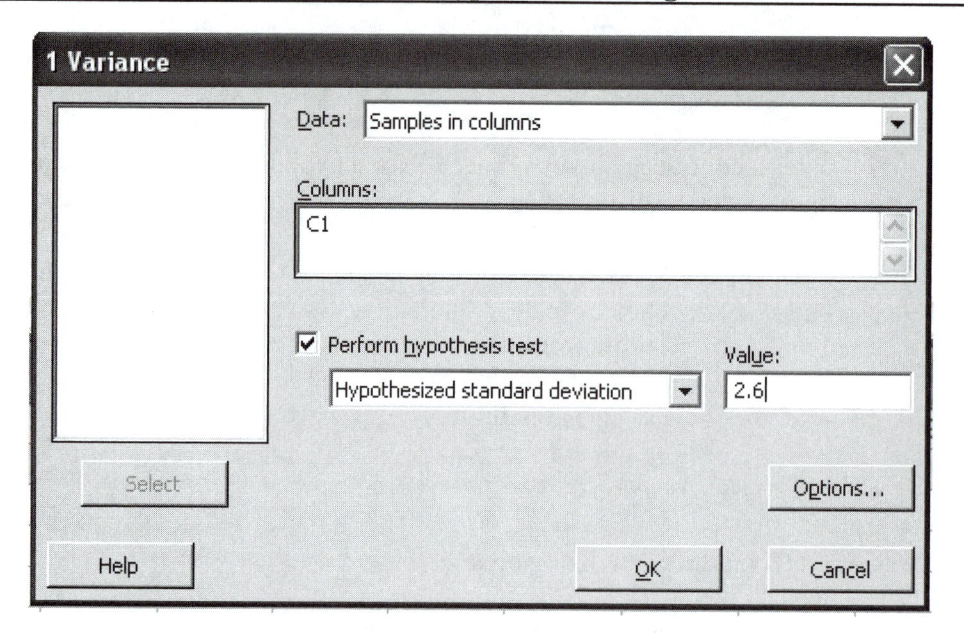

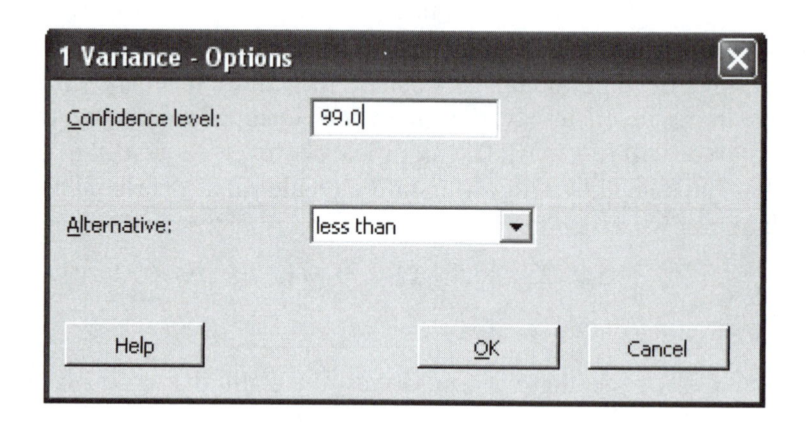

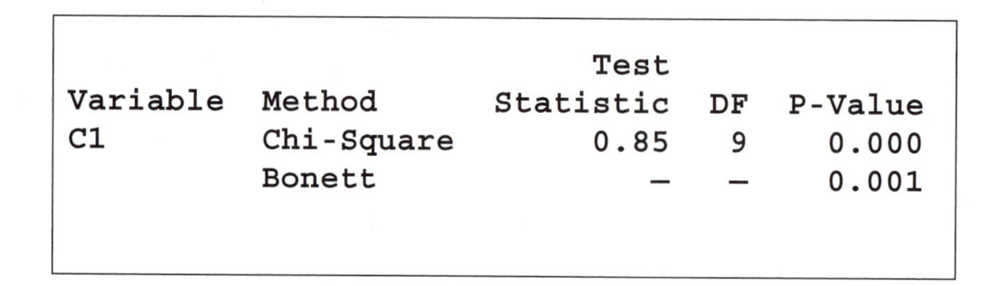

Variable	Method	Test Statistic	DF	P-Value
C1	Chi-Square	0.85	9	0.000
	Bonett	–	–	0.001

The Minitab results include the test statistic of $\chi^2 = 0.85$ and a P-value of 0.000. (The actual P-value is 0.000290, but it becomes 0.000 when rounded to three decimal places. Because the P-value is low, we reject the null hypothesis and conclude that there is sufficient evidence to support the claim that the population standard deviation σ is less than 2.6 in. (The bottom line includes results from application of Bonnett's method, which is not discussed in the textbook.)

8-5 Testing Hypotheses with Simulations

Sections 8-1, 8-2, and 8-3 of this manual/workbook have all focused on using Minitab for hypothesis tests using the *P*-value approach. Another very different approach is to use *simulations*. Let's illustrate the simulation technique with an example.

Consider testing the claim that the population of body temperatures of healthy adults has a mean less than 98.6°F. That is, consider the claim that $\mu < 98.6$°F. We will use sample data consisting of these 12 values:

98.0 97.5 98.6 98.8 98.0 98.5 98.6 99.4 98.4 98.7 98.6 97.6

For these 12 values, $n = 12$, $\bar{x} = 98.39167$, and $s = 0.53506$. The key question is this:

> If the population mean body temperature is really 98.6, then *how likely* is it that we would get a sample mean of 98.39167, given that the population has a normal distribution and the sample size is 12?

If the probability of getting a sample mean such as 98.39167 is very small, then that suggests that the sample values are not the result of chance random fluctuation. If the probability is high, then we can accept random chance as an explanation for the discrepancy between the sample mean of 98.39167 and the assumed mean of 98.6. What we need is some way of determining the likelihood of getting a sample mean such as 98.39167. That is the precise role of *P*-values in the *P*-value approach to hypothesis testing. However, there is another approach. Minitab and many other software packages are capable of generating random results from a variety of different populations. Here is how Minitab can be used: Determine the likelihood of getting a sample mean of 98.39167 by randomly generating different samples from a population that is normally distributed with the claimed mean of 98.6. For the standard deviation, we will use the best available information: the value of $s = 0.53506$ obtained from the sample.

Minitab Procedure for Testing Hypotheses with Simulations

1. Identify the values of the sample size *n*, the sample standard deviation *s*, and the claimed value of the population mean.

2. Click on **Calc**, then select **Random Data**.

3. Click on **Normal.**

4. Generate a sample randomly selected from a normally distributed population with the claimed mean. When making the required entries in the dialog box, use the *claimed* mean, the sample size *n*, and the sample standard deviation *s*.

5. Continue to generate similar samples until it becomes clear that the given sample mean is or is not likely. (Here is one criterion: The given sample mean is *unlikely* if its value or more extreme values occur 5% of the time or less.) If it is unlikely, reject the claimed mean. If it is likely, fail to reject the claimed mean.

For example, here are 20 results obtained from the random generation of samples of size 12 from a normally distributed population with a mean of 98.6 and a standard deviation of 0.53506:

Sample Means

98.754	**98.374**	**98.332**	98.638	98.513
98.551	98.566	98.760	**98.332**	98.603
98.407	98.640	98.655	98.408	98.802
98.505	98.699	98.609	98.206	98.582

Examining the 20 sample means, we see that three of them (displayed in bold) are 98.39167 or lower. Because 3 of the 20 results (or 15%) are at least as extreme as the sample mean of 98.39167, we see that a sample mean such as 98.39167 is *common* for these circumstances. This suggests that a sample mean of 98.39167 is not *significantly* different from the assumed mean of 98.6. We would feel more confident in this conclusion if we had more sample results, so we could continue to randomly generate simulated samples until we feel quite confident in our thinking that a sample mean such as 98.39167 is not an unusual result. It can easily occur as the result of chance random variation. We therefore fail to reject the null hypothesis that the mean equals 98.6. There is not sufficient evidence to support the claim that the mean is less than 98.6.

CHAPTER 8 EXPERIMENTS: Hypothesis Testing

Experiments 8–1 through 8–4 involve claims about proportions. Use Minitab with the normal approximation procedure for the following.

8–1. ***Reporting Income*** In a Pew Research Center poll of 745 randomly selected adults, 589 said that it is morally wrong to not report all income on tax returns. Use a 0.01 significance level to test the claim that 75% of adults say that it is morally wrong to not report all income on tax returns.

Test statistic: __________ *P*–value: __________

Conclusion in your own words: ____

____

8–2. ***Voting for the Winner*** In a presidential election, 308 out of 611 voters surveyed said that they voted for the candidate who won (based on data from ICR Survey Research Group). Use a 0.01 significance level to test the claim that among all voters, the percentage believing that they voted for the winning candidate is equal to 43%, which is the actual percentage of votes for the winning candidate.

Test statistic: __________ *P*–value: __________

Conclusion in your own words: ____

____

8–3. ***Tennis Instant Replay*** The Hawk-Eye electronic system is used in tennis for displaying an instant replay that shows whether a ball is in bounds or out of bounds, so players can challenge calls made by referees. In the most recent U.S. Open (as of this writing), singles players made 611 challenges and 172 of them were successful with the call overturned. Use a 0.01 significance level to test the claim that fewer than 1/3 of the challenges are successful.

Test statistic: __________ *P*–value: __________

Conclusion in your own words: ____

____

8–4. ***Screening for Marijuana Usage*** The company Drug Test Success provides a "1-Panel-THC" test for marijuana usage. Among 300 tested subjects, results from 27 subjects were wrong (either a false positive or a false negative). Use a 0.05 significance level to test the claim that less than 10% of the test results are wrong.

Test statistic: __________ *P*–value: __________

Conclusion in your own words: ____

____

8-5. ***Exact Procedure and Normal Approximation*** Repeat Experiment 8-4 by using Minitab's exact procedure. (In the **Options** dialog box, do not click on the box next to **Use test and interval based on normal distribution**.)

Test statistic: __________ *P*–value: __________

Conclusion in your own words: ____

____

Compare the exact results to those found by using the normal approximation.

____

____

Experiments 8–6 through 8–9 use Minitab with the original list of sample data for testing the given claim about a population mean.

8–6. ***Earthquake magnitudes*** Use the earthquake magnitudes listed in Data Set 16 in Appendix B and test the claim that the population of earthquakes has a mean magnitude greater than 1.00. Use a 0.05 significance level.

Test statistic: __________ *P*–value: __________

Conclusion in your own words: ____

____

8–7. ***Blood Pressure*** Use the systolic blood pressure measurements for females lised in Data Set 1 in Appendix B and test the claim that the female population has a mean sytolic blood pressure level less than 120.0 mm Hg. Use a 0.05 significance level.

Test statistic: __________ *P*–value: __________

Conclusion in your own words: ____

____

8–8. ***College Weights*** Use the September weights of males in Data Set 4 from Appendix B and test the claim that male college students have a mean weight that is less than the 83 kg mean weight of males in the general population. Use a 0.01 significance level.

Test statistic: __________ *P*–value: __________

Conclusion in your own words: ____

____

8-9. ***Power Supply*** Data Set 18 in Appendix B lists measured voltage amounts supplied directly to the author's home. The Central Hudson power supply company states that it has a target power supply of 120 volts. Using those home voltage amounts, test the claim that the mean is 120 volts. Use a 0.01 significance level.

Test statistic: __________ *P*–value: __________

Conclusion in your own words: ____

____

In Experiments 8–10 through 8–13, test the claim about a population mean.

8–10. ***Cigarette Tar*** A simple random sample of 25 filtered 100 mm cigarettes is obtained, and the tar content of each cigarette is measured. The sample has a mean of 13.2 mg and a standard deviation of 3.7 mg (based on Data Set 10 in Appendix B). Use a 0.05 significance level to test the claim that the mean tar content of filtered 100 mm cigarettes is less than 21.1 mg, which is the mean for unfiltered king size cigarettes.

Test statistic: __________ *P*–value: __________

Conclusion in your own words: ____

____

8–11. ***Is the Diet Practical?*** When 40 people used the Weight Watchers diet for one year, their mean weight *loss* was 3.0 lb and the standard deviation was 4.9 lb (based on data from "Comparison of the Atkins, Ornish, Weight Watchers, and Zone Diets for Weight Loss and Heart Disease Reduction," by Dansinger, et al., *Journal of the American Medical Association,* Vol. 293, No. 1). Use a 0.01 significance level to test the claim that the mean weight loss is greater than 0 lb.

Test statistic: __________ *P*–value: __________

Conclusion in your own words: ____

____

Based on these results, does the diet appear to be effective? Does the diet appear to have practical significance? Explain.

____

____

8–12. ***Weights of Pennies*** The U. S. Mint has a specification that pennies have a mean weight of 2.5 g. Data Set 21 in Appendix B lists the weights (in grams) of 37 pennies manufactured after 1983. Those pennies have a mean weight of 2.49910 g and a standard deviation of 0.01648 g. Use a 0.05 significance level to test the claim that this sample is from a population with a mean weight equal to 2.5 g.

Test statistic: __________ *P*–value: __________

Conclusion in your own words: ____

____

8–13. ***Analysis of Pennies*** In an analysis investigating the usefulness of pennies, the cents portions of 100 randomly selected credit card charges are recorded. The sample has a mean of 47.6 cents and a standard deviation of 33.5 cents. If the amounts from 0 cents to 99 cents are all equally likely, the mean is expected to be 49.5 cents. Use a 0.01 significance level to test the claim that the sample is from a population with a mean equal to 49.5 cents.

Test statistic: __________ *P*–value: __________

Conclusion in your own words: ____

In Experiments 8–14 through 8–17, use Minitab to test the claim about a standard deviation or variance.

8–14. ***Ages of Race Car Drivers*** Listed below are the ages (years) of randomly selected race car drivers (based on data reported in *USA Today*). Most people in the general population have ages that vary between 0 and 90 years, so use of the range rule of thumb suggests that ages in the general population have a standard deviation of 22.5 years. Use a 0.01 significance level to test the claim that the standard deviation of ages of all race car drivers is less than 22.5 years.

32 32 33 33 41 29 38 32 33 23 27 45 52 29 25

Test statistic: __________ *P*–value: __________

Conclusion in your own words: ____

____

8–15. ***Highway Speeds*** Listed below are speeds (mi/h) measured from southbound traffic on I-280 near Cupertino, California (based on data from SigAlert). This simple random sample was obtained at 3:30 PM on a weekday. Use a 0.05 significance level to test the claim of the highway engineer that the standard deviation of speeds is equal to 5.0 mi/h.

62 61 61 57 61 54 59 58 59 69 60 67

Test statistic: __________ *P*–value: __________

Conclusion in your own words: ____

____

8-16. ***Aircraft Altimeters*** The Skytek Avionics company uses a new production method to manufacture aircraft altimeters. A simple random sample of new altimeters resulted in errors listed below. Use a 0.05 level of significance to test the claim that the new production method has errors with a standard deviation greater than 32.2 ft, which was the standard deviation for the old production method. If it appears that the standard deviation is greater, does the new production method appear to be better or worse than the old method? Should the company take any action?

–42 78 –22 –72 –45 15 17 51 –5 –53 –9 –109

Test statistic: __________ *P*–value: __________

Conclusion in your own words: ____

____

8–17. ***IQ of Professional Pilots*** The Wechsler IQ test is designed so that the mean is 100 and the standard deviation is 15 for the population of normal adults. Listed below are IQ scores of randomly selected professional pilots. It is claimed that because professional pilots are a more homogeneous group than the general population, they have IQ scores with a standard deviation less than 15. Test that claim using a 0.05 significance level.

121 116 115 121 116 107 127 98 116 101 130 114

Test statistic: __________ *P*–value: __________

Conclusion in your own words: ____

____

Simulations *The following experiments involve the simulation approach to hypothesis testing.*

8–18. ***Hypothesis Testing with Simulations*** Suppose that we want to use a 0.05 significance level to test the claim that $p > 0.5$ and we have sample data summarized by $n = 100$ and $x = 70$. Use the following Minitab procedure for using a simulation method for testing the claim that $p > 0.5$.

We assume in the null hypothesis that $p = 0.5$. Working under the assumption that $p = 0.5$, we will generate different samples of size $n = 100$ until we have a sense for the likelihood of getting 70 or more successes. Begin by generating a sample of size 100 as follows: Select **Calc/Random Data/Integer**. In the dialog box, enter 100 for the number of rows, enter C1–C10 for the columns, then enter a minimum of 0 and a maximum of 1. After clicking **OK**, the result will be 10 columns, where each column represents a sample of 100 subjects with a probability of 0.5 of getting a 1 in each cell. Letting 1 represent a success, use **Stat/Display Descriptive Statistics** and find the number of samples having a sample proportion of 0.70 or greater. Enter that result here: __________

What percentage of the simulated samples have a sample proportion of 0.70 or greater? Enter that result here: _________

What does the preceding result suggest about the likelihood of this event:
"When the population proportion p is really equal to 0.5, a sample of size 100 is randomly selected and the sample proportion is 0.70 or greater."

____

Can we conclude that the probability of a sample proportion of 0.70 or greater is less than or equal to the significance level of 0.05? __________

What is the final conclusion? ____

____

8–19. ***Hypothesis Testing with Simulations*** Experiment 8–18 used a simulation approach to test a claim. Use a simulation approach for conducting the hypothesis test described in Experiment 8–2. Describe the procedure, results, and conclusions.

____

____

____

____

8–20. ***Hypothesis Testing with Simulations*** Refer to Experiment 8–10 in this manual/work–book. We want to test the claim that $\mu < 21.1$, and the given sample data can be summarized with these statistics: $n = 25$, $\bar{x} = 13.2$, and $s = 3.7$. Instead of conducting a formal hypothesis test, we will now consider another way of determining whether the sample mean of 13.2 is significantly less than the claimed value of 21.1. We will use a significance level of 0.05 with this criterion:

> *The sample mean of 13.2 is significantly less than 21.1 if there is a 5% chance (or less) that the following event occurs: A sample mean of 13.2 or less is obtained when selecting a random sample of $n = 25$ values from a normally distributed population with mean $\mu = 21.1$ and standard deviation $\sigma = 3.7$.*

Use **Calc/Random Data/Normal** to randomly generate a sample of 25 values from a normally distributed population with the assumed mean of 21.1 and a standard deviation of 3.7. Find the mean of the generated sample and record it in the space below. Then generate another sample, and another sample, and continue until you have enough sample means to determine how often a result such as $\bar{x} = 13.2$ (or less) will occur.

____

____

Enter the number of generated samples:________________

How many of the generated samples have a sample mean less than 21.1? ____________

What is the proportion of trials in which the sample mean was 21.1 or less? __________

Based on these results, what do you conclude about the claim that $\mu < 21.1$? Explain.

____

9

Inferences from Two Samples

9–1 Minitab's Assistant Feature

Minitab introduced its new *Assistant* feature in Release 16. If you click on the main menu item of **Assistant,** you get a submenu that includes the option of **Hypothesis Tests.** If you select Hypothesis Tests, you are given options of comparing one sample with a target (as in Chapter 8), or comparing two samples with each other (as in this Chapter), or comparing more than two samples. Shown below are the choices available under the comparison of two samples with each other, and those choices include the topics covered in this chapter. An advantage of using the Assistant feature is that you are provided with some guidance in choosing the correct method. Also, after entering information about the sample, you will get different screens of results that provide much information that is helpful in interpreting your results. The following sections of this manual/workbook describe the use of Minitab's basic functions, but consider using the *Assistant* feature in place of those basic methods.

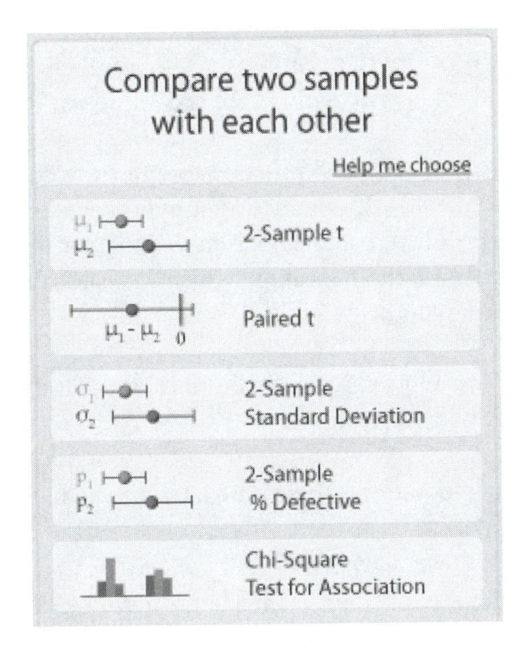

9-2 Two Proportions

The textbook makes the point that the section discussing inferences involving two proportions is one of the most important sections in the book because the main objective is to provide methods for dealing with two sample proportions — a situation that is very common in real applications.

To use Minitab, we must first identify the number of successes x_1 and the sample size n_1 for the first sample, and identify x_2 and n_2 for the second sample. Sample data often consist of sample proportions or percentages instead of the actual numbers of successes, so we must know how to determine the number of successes. From $\hat{p}_1 = x_1/n_1$, we know that $x_1 = n_1 \cdot \hat{p}_1$ so that x_1 can be found by multiplying the sample size for the first sample by the sample proportion expressed in decimal form. For example, if 35.2% of 1200 interviewed subjects answer "yes" to a question, we have $n = 1200$ and $x = 1200 \cdot 0.352 = 422.4$, which we round to 422.

After identifying n_1, x_1, n_2, and x_2, proceed with Minitab as follows.

Minitab Procedure for Inferences (Hypothesis Tests and Confidence Intervals) for Two Proportions

To conduct hypothesis tests or construct confidence intervals for two population proportions, use the following procedure.

1. For each sample, find the sample size n and the number of successes x.

2. Select **Stat**, then **Basic Statistics**, then **2 Proportions**.

3. In the dialog box, select the option of "Summarized data" and enter the number of successes (or "events") for each of the two samples, and enter the sample size (or "trials") for each of the two samples. (The numbers of "Events" are the values of x_1 and x_2. The numbers of "Trials" are the values of n_1 and n_2.)

4. Click on the **Options** bar and proceed to enter the confidence level (enter 95 for a 0.05 significance level). If conducting a hypothesis test, enter the claimed value of the difference (usually 0), and select the format for the alternative hypothesis. (See the Minitab screen on the next page where "greater than" is selected for the format of the alternative hypothesis.)

5. While still in the Options dialog box, refer to the box with the label of "Use pooled estimate of p for test."

   - If testing a hypothesis, click on that box (because you want to use a pooled estimate of p).

   - If constructing a confidence interval, do *not* click on that box (because you do not want to use a pooled estimate of p).

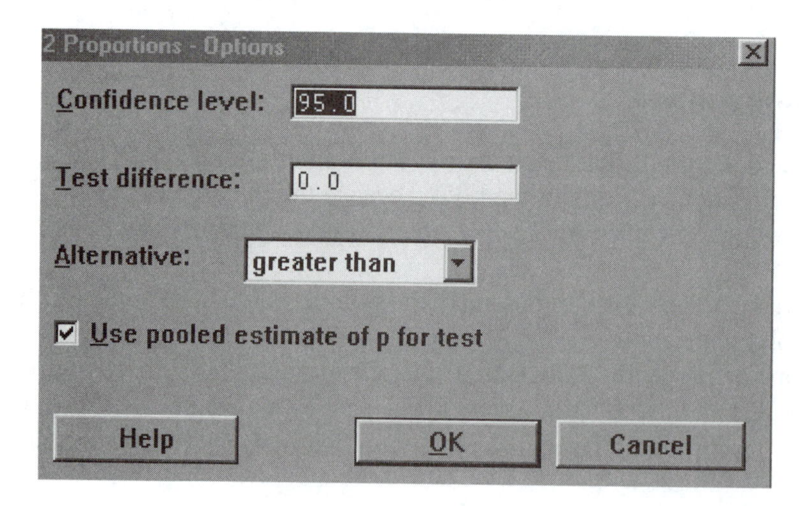

As an illustration, consider the following sample data from Example 1 in Section 9-2 of the textbook:

> **Large Denominations Less Likely to be Spent?** The table below lists sample results from a study conducted to determine whether people are less likely to spend money when it is in the form of larger denominations. Use a 0.05 significance level to test the claim that "money in a large denomination is less likely to be spent relative to an equivalent amount in many smaller denominations." That is, test the claim that the proportion of people who spend a $1 bill is less than the proportion who spend 4 quarters.

	Group 1 Subjects Given $1 Bill	*Group 2* Subjects Given 4 Quarters
Spent the money	$x_1 = 12$	$x_2 = 27$
Subjects in Group	$n_1 = 46$	$n_2 = 43$

We can now proceed to use the above Minitab procedure. After selecting **Stat**, **Basic Statistics**, then **2 Proportions**, we make the required entries in the dialog boxes as shown below.

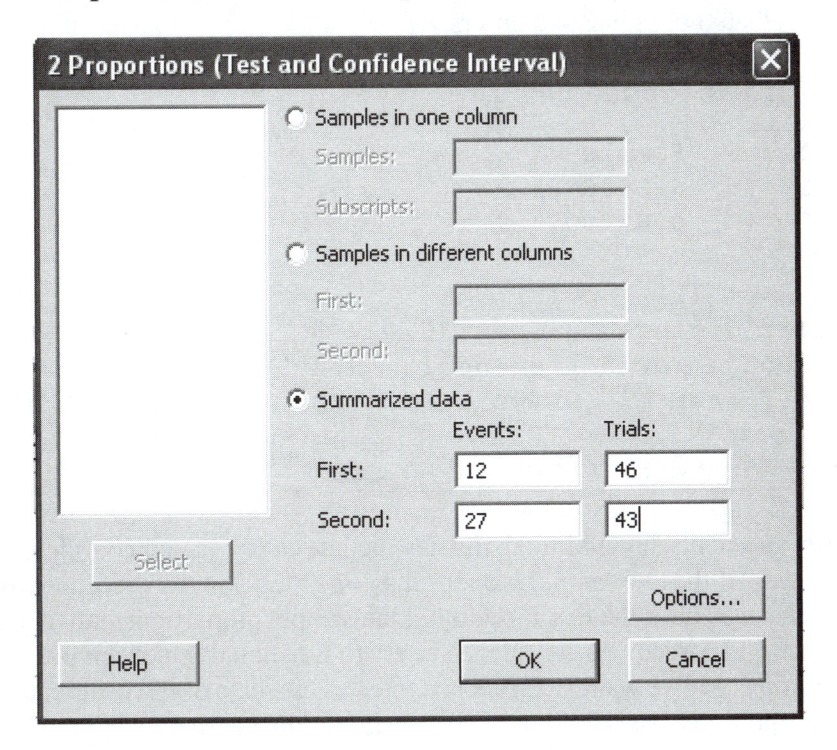

The claim of a lower spending rate for the population given the $1 bill is the claim that $p_1 < p_2$, so we select the option of "less than" for the alternative hypothesis, as shown below. Also, because we are conducting a hypothesis test, the two sample estimates of the two population proportions are *pooled,* so we check the box as shown below.

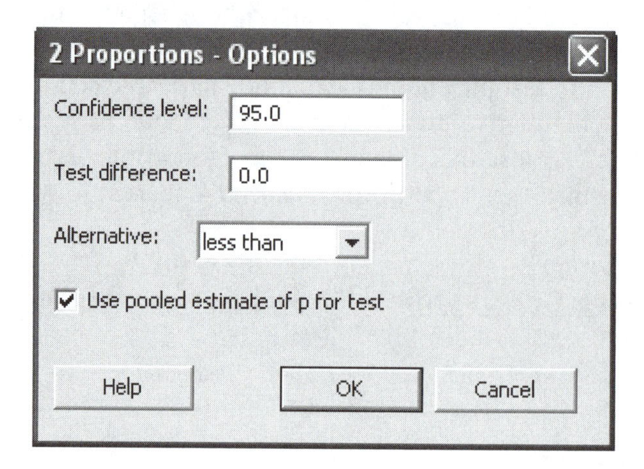

After clicking on **OK** in both dialog boxes, the Minitab results will be displayed. Shown below are the results from the preceding example and dialog box entries. The results include the test statistic of $z = -3.49$ and the *P*–value of 0.000. Because the *P*–value of 0.000 is less than the significance level of 0.05, we reject the null hypothesis and support the claim that those given $1 spend at a rate less than those given 4 quarters..

Test and CI for Two Proportions

```
Sample    X   N   Sample p
1        12  46   0.260870
2        27  43   0.627907

Difference = p (1) - p (2)
Estimate for difference:  -0.367037
95% upper bound for difference:  -0.205664
Test for difference = 0 (vs < 0):  Z = -3.49  P-Value = 0.000

Fisher's exact test: P-Value = 0.000
```

Confidence Intervals: The above Minitab display includes a one–sided confidence interval, but it is based on *pooling* of the two sample proportions. (See Step 5 in the preceding Minitab procedure. Note that we check the box for pooling the sample proportions only if we are testing a hypothesis; if we are constructing a confidence interval, we should *not* check the box for pooling the sample proportions.) If we want a confidence interval estimate of the difference between the two population proportions, we should use "not equal" for the alternative hypothesis and we should *not* check the box for pooling. Minitab provides this 90% confidence interval estimate of the difference between the two population proportions: (-0.528410, -0.205664). This can also be expressed as $-0.528 < p_1 - p_2 < -0.206$ (rounded).

9-3 Two Means: Independent Samples

The textbook notes that two samples are **independent** if the sample values selected from one population are not related to or somehow paired or matched with the sample values selected from the other population. If there is some relationship so that each value in one sample is paired with a corresponding value in the other sample, the samples are **dependent.** Dependent samples are often referred to as **matched pairs,** or **paired samples.**

When testing a claim about the means of two independent samples, or when constructing a confidence interval estimate of the difference between the means of two independent samples, the textbook describes procedures based on the requirement that the two population standard deviations σ_1 and σ_2 are *not known*, and there is no assumption that $\sigma_1 = \sigma_2$. We focus on the first of the following three cases, and the other two cases are discussed briefly:

1. σ_1 and σ_2 are not known and are not assumed to be equal.
2. σ_1 and σ_2 are known.
3. It is assumed that $\sigma_1 = \sigma_2$.

The first of these three cases is the main focus of Section 9-3 in the textbook, and Minitab handles that first case but, instead of using "the smaller of $n_1 - 1$ and $n_2 - 1$" for the number of degrees of freedom, Minitab calculates the number of degrees of freedom using Formula 9-1 in Section 9-3 of *Elementary Statistics,* 12th edition.

Minitab also provides the option of assuming that $\sigma_1 = \sigma_2$, but there is no option for assuming that σ_1 and σ_2 are *known*. That is, Minitab allows us to deal with cases 1 and 3 listed above, but not case 2. We describe the Minitab procedure, then consider an example.

Minitab Procedure for Inferences (Hypothesis Tests and Confidence Intervals) about Two Means: Independent Samples

Use the following procedure for both hypothesis tests and confidence intervals.

1. Identify the values of the summary statistics, or enter the original sample values in columns C1 and C2 (or open a Minitab worksheet with columns containing the lists of sample values).

2. Click on the main menu item of **Stat**, select **Basic Statistics**, then **2-Sample t**.

3. Proceed to make the entries in the dialog box.

    - Do *not* check the box labeled "Assume equal variances" (unless there is a sound justification for assuming that $\sigma_1 = \sigma_2$). Be sure that the box remains unchecked. We are not assuming equal variances.

    - Click on the **Options** bar and enter the confidence level (enter 95 for a 0.05 significance level). If conducting a hypothesis test, enter the claimed value of the difference (usually 0), and also select the format for the alternative hypothesis. Click **OK** when finished.

As an illustration, consider the following example.

Are Men and Women Equal Talkers? A headline in *USA Today* proclaimed that "Men, women are equal talkers." That headline referred to a study of the numbers of words that samples of men and women spoke in a day. Given below are the results from the study, which are included in Data Set 17 in Appendix B (based on "Are Women Really More Talkative Than Men?" by Mehl, et al, *Science*, Vol. 317, No. 5834). Use a 0.05 significance level to test the claim that men and women speak the same mean number of words in a day. Does there appear to be a difference?

Number of Words Spoken in a Day

Men	Women
$n_1 = 186$	$n_2 = 210$
$\bar{x}_1 = 15,668.5$	$\bar{x}_2 = 16,215.0$
$s_1 = 8632.5$	$s_2 = 7301.2$

Using Minitab with the preceding procedure, we enter the summary statistics as shown below.

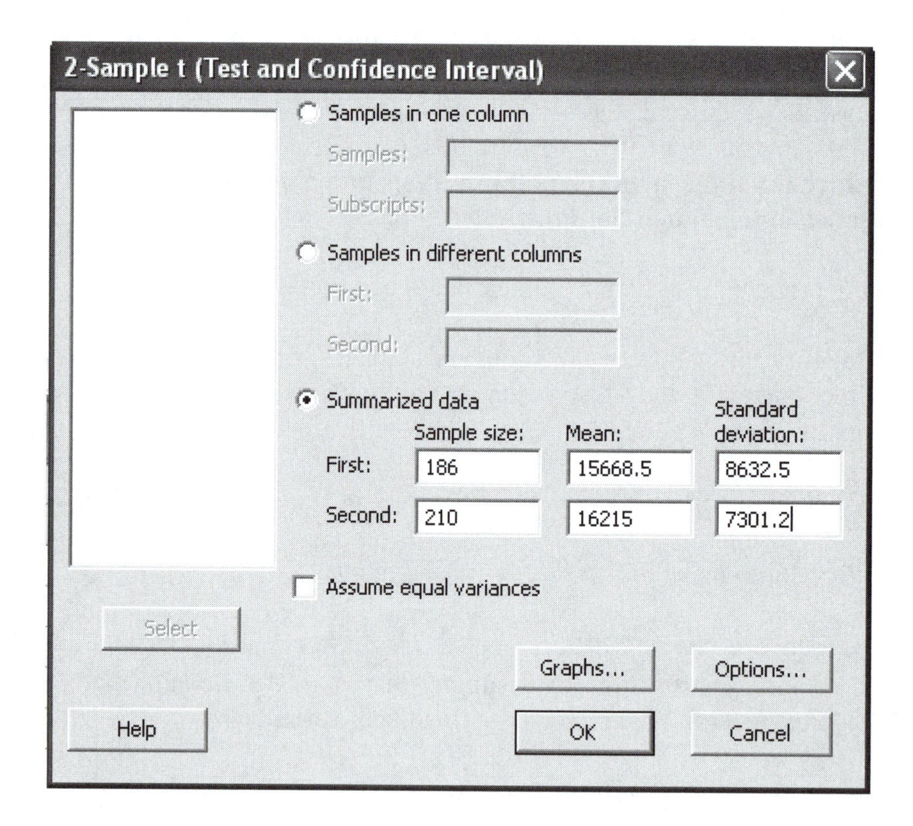

Click on the **Options** button, then make the selections shown in the following dialog box.

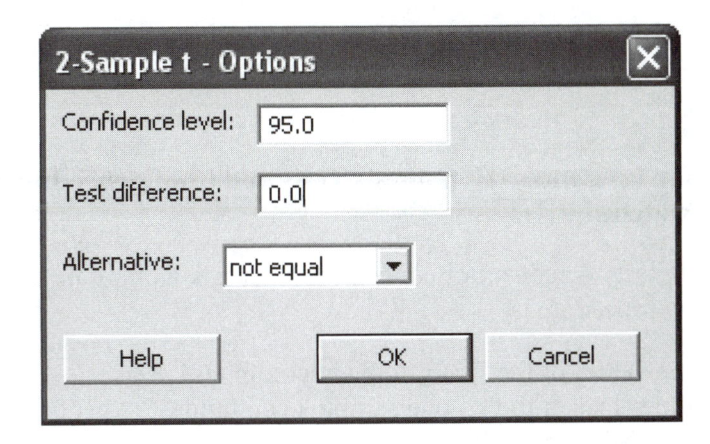

The Minitab results are as shown below. The test statistic is $t = -0.68$ and the P–value is 0.500, so we fail to reject the null hypothesis. There is not sufficient evidence to reject the claim that men and women speak the same mean number of words in a day.

Two-Sample T-Test and CI

```
Sample    N   Mean  StDev  SE Mean
1       186  15669   8633      633
2       210  16215   7301      504

Difference = mu (1) - mu (2)
Estimate for difference:  -547
95% CI for difference:  (-2137, 1044)
T-Test of difference = 0 (vs not =): T-Value = -0.68  P-Value = 0.500
DF = 364
```

Confidence Interval If we wanted to construct a confidence interval for the difference between the mean number of words spoken by men and the mean number of words spoken by women in day, the results above show that the 95% confidence interval is $-2137 < \mu_1 - \mu_2 < 1044$. Because that confidence interval includes 0, there does not appear to be a significant difference between the two means.

9-4 Matched Pairs

The textbook describes methods for testing hypotheses and constructing confidence interval estimates of the differences between samples consisting of *matched pairs*. Here is the Minitab procedure:

Minitab Procedure for Inferences (Hypothesis Tests and Confidence Intervals) about Two Means: Matched Pairs

To conduct hypothesis tests or construct confidence intervals using data consisting of matched pairs, use the following procedure.

1. Enter the values of the first sample in column C1.
   Enter the values of the second sample in column C2
   (or open a Minitab worksheet that includes columns of matched pairs of sample data).

2. Select **Stat**, **Basic Statistics**, **Paired t**.

3. Make these entries and selections in the dialog box that pops up:

   - Enter C1 (or the column label) in the box labeled "First sample.
   - Enter C2 (or the column label) in the box labeled "Second sample."
   - Click the **Options** bar and proceed to enter the confidence level (enter 95 for a 0.05 significance level). If testing a claim about the mean of the differences between the paired data, enter the claimed value of the difference μ_d, and select the form of the alternative hypothesis. Click **OK**, then click **OK** for the main dialog box.

See the following example.

Hypothesis Test of Claimed Freshman Weight Gain Listed below are samples of measured weights of college students in September and April of their freshman year. Use the sample data with a 0.05 significance level to test the claim that for the population of students, the mean change in weight from September to April is equal to 0 kg.

Weight (kg) Measurements of Students in Their Freshman Year

April weight	66	52	68	69	71
September weight	67	53	64	71	70

Shown on the next page are the dialog boxes and the Minitab results.

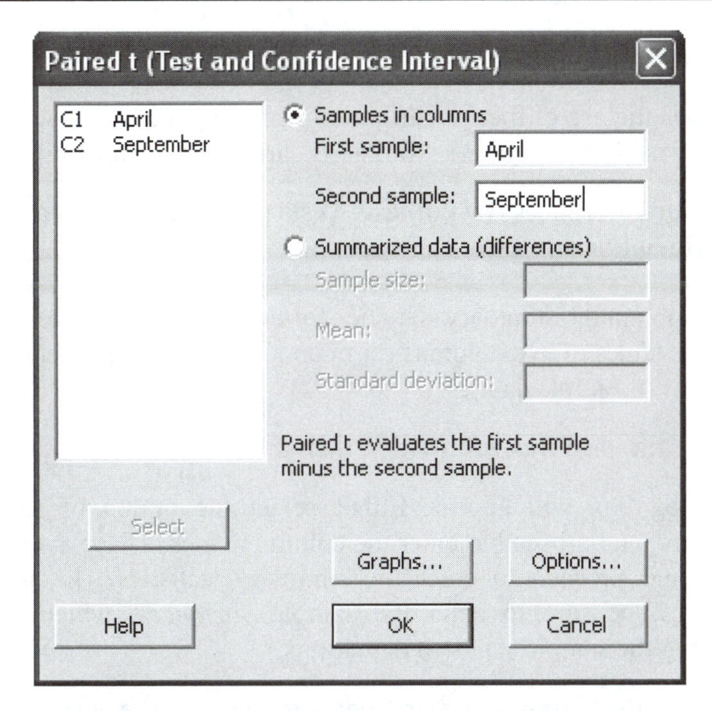

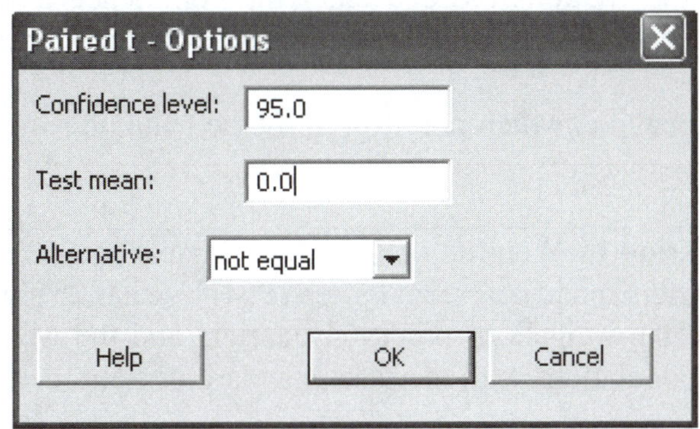

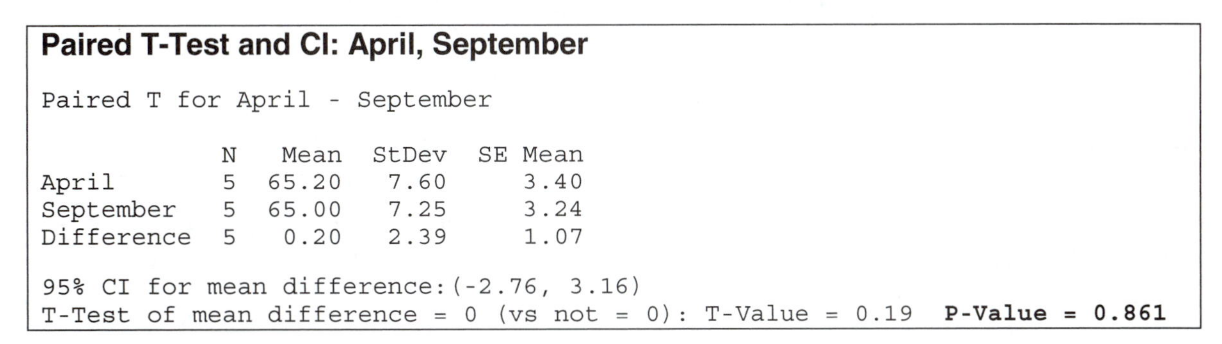

Paired T-Test and CI: April, September

```
Paired T for April - September

             N   Mean  StDev  SE Mean
April        5  65.20   7.60     3.40
September    5  65.00   7.25     3.24
Difference   5   0.20   2.39     1.07

95% CI for mean difference:(-2.76, 3.16)
T-Test of mean difference = 0 (vs not = 0): T-Value = 0.19  P-Value = 0.861
```

The P-value of 0.861 is greater than the significance level of 0.05, so we fail to reject the null hypothesis that the mean of the differences between the April weights and the September weights is equal to 0 kg. There does not appear to be a significant difference.

Confidence Interval The above Minitab results show that the 95% confidence interval estimate of the mean of the differences is $-2.8 \text{ kg} < \mu_d < 3.2 \text{ kg}$ (rounded). Because the confidence interval includes 0 kg, there does not appear to be a significant difference.

9-5 Two Variances

The textbook describes the use of the *F* distribution in testing a claim that two populations have the same variance (or standard deviation). Minitab is capable of conducting such an *F* test.

Minitab Procedure for Inferences (Hypothesis Tests and Confidence Intervals) for Two Standard Deviations or Two Variances

1. Either obtain the summary statistics for each sample, or enter the individual sample values in two columns (or open a Minitab worksheet containing the columns of sample data).
2. Select **Stat**, then **Basic Statistics**, then **2 Variances**.
3. A dialog box will appear. Either select the option of "Samples in different columns" and proceed to enter the column names or labels, or select "Summarized data" and proceed to enter the summary statistics. (If entering the summary statistics, be sure to enter the sample *variances*, which can be obtained by squaring the sample standard deviations.)
4. Click on the **Options** bar and enter the confidence level. (Enter 95 for a 0.05 significance level.) Click **OK**, then click **OK** in the main dialog box.

Consider the following example, which is followed by the Minitab results obtained by using the above procedure.

Comparing Variation in Weights of Quarters Sample statistics are listed below for weights of quarters made before and after 1964. Use a 0.05 significance level to test the claim that the weights of pre-1964 quarters and the weights of post-1964 quarters are from populations with the same standard deviation.

Pre-1964 Quarters	Post-1964 Quarters
$n = 40$	$n = 40$
$s = 0.08700$ g	$s = 0.06194$ g

Minitab Results

Method	DF1	DF2	Test Statistic	P-Value
F Test (normal)	**39**	**39**	**1.97**	**0.037**
Levene's Test (any continuous)	1	78	4.22	0.043

For the *F*-test method described in Section 9-5 of the textbook, the Minitab results include the *P*-value of 0.037, which is less than the significance level of 0.05. We therefore reject the null hypothesis that the two populations of quarters have weights with the same standard deviation.

CHAPTER 9 EXPERIMENTS: Inferences from Two Samples

9-1. ***Drug Use in College*** In a 1993 survey of 560 college students, 171 said that they used illegal drugs during the previous year. In a recent survey of 720 college students, 263 said that they used illegal drugs during the previous year (based on data from the National Center for Addiction and Substance Abuse at Columbia University). Use a 0.05 significance level to test the claim that the proportion of college students using illegal drugs in 1993 was less than it is now.

Test statistic: __________ *P*–value: __________

Conclusion in your own words: ____

____

9-2. ***Drug Use in College*** Using the sample data from Experiment 9-1, construct the confidence interval corresponding to the hypothesis test conducted with a 0.05 significance level. What conclusion does the confidence interval suggest?

____

____

9-3. ***Are Seat Belts Effective?*** A simple random sample of front-seat occupants involved in car crashes is obtained. Among 2823 occupants not wearing seatbelts, 31 were killed. Among 7765 occupants wearing seatbelts, 16 were killed (based on data from "Who Wants Airbags?" by Meyer and Finney, *Chance*, Vol. 18, No. 2). Construct a 90% confidence interval estimate of the difference between the fatality rates for those not wearing seat belts and those wearing seat belts. What does the result suggest about the effectiveness of seat belts?

____

____

9-4. ***Are Seat Belts Effective?*** Use the sample data in Experiment 9-3 with a 0.05 significance level to test the claim that the fatality rate is higher for those not wearing seat belts.

Test statistic: __________ *P*–value: __________

Conclusion in your own words: ____

____

____

9-5. ***Morality and Marriage*** A Pew Research Center poll asked randomly selected subjects if they agreed with the statement that "It is morally wrong for married people to have an affair." Among the 386 women surveyed, 347 agreed with the statement. Among the 359 men surveyed, 305 agreed with the statement. Use a 0.05 significance level to test the claim that the percentage of women who agree is different from the percentage of men who agree. Does there appear to be a difference in the way women and men feel about this issue?

Test statistic: __________ *P*–value: __________

Conclusion in your own words: ____

____

9-6. ***Morality and Marriage*** Using the sample data from Experiment 9-5, construct the confidence interval corresponding to the hypothesis test conducted with a 0.05 significance level. What conclusion does the confidence interval suggest?

____

____

9-7. ***Cardiac Arrest at Day and Night*** A study investigated survival rates for in-hospital patients who suffered cardiac arrest. Among 58,593 patients who had cardiac arrest during the day, 11,604 survived and were discharged. Among 28,155 patients who suffered cardiac arrest at night, 4139 survived and were discharged (based on data from "Survival from In-Hospital Cardiac Arrest During Nights and Weekends," by Peberdy et al., *Journal of the American Medical Association,* Vol. 299, No. 7). Use a 0.01 significance level to test the claim that the survival rates are the same for day and night.

Test statistic: __________ *P*–value: __________

Conclusion in your own words: ____

____

9-8. ***Cardiac Arrest at Day and Night*** Using the sample data from Experiment 9-7, construct the confidence interval corresponding to the hypothesis test conducted with a 0.01 significance level. What conclusion does the confidence interval suggest?

____

____

In Experiments 9–9 through 9–20, assume that the two samples are independent simple random samples selected from normally distributed populations. Do not assume that the population standard deviations are equal.

9-9. ***Hypothesis Test of Effectiveness of Humidity in Treating Croup*** Treatment In a randomized controlled trial conducted with children suffering from viral croup, 46 children were treated with low humidity while 46 other children were treated with high humidity. Researchers used the Westley Croup Score to assess the results after one hour. The low humidity group had a mean score of 0.98 with a standard deviation of 1.22 while the high humidity group had a mean score of 1.09 with a standard deviation of 1.11 (based on data from "Controlled Delivery of High vs Low Humidity vs Mist Therapy for Croup Emergency Departments," by Scolnik et al, *Journal of the American Medical Association,* Vol. 295, No. 11). Use a 0.05 significance level to test the claim that the two groups are from populations with the same mean. What does the result suggest about the common treatment of humidity?

Test statistic: __________ *P*–value: __________

Conclusion in your own words: ____

____

9-10. ***Confidence Interval for Effectiveness of Humidity in Treating Croup*** Use the sample data given in Experiment 9-9 and construct a 95% confidence interval estimate of the difference between the mean Westley Croup Score of children treated with low humidity and the mean score of children treated with high humidity. What does the confidence interval suggest about humidity as a treatment for croup?

____

____

9-11. ***Confidence Interval for Cigarette Tar*** The mean tar content of a simple random sample of 25 unfiltered king-size cigarettes is 21.1 mg, with a standard deviation of 3.2 mg. The mean tar content of a simple random sample of 25 filtered 100 mm cigarettes is 13.2 mg with a standard deviation of 3.7 mg (based on data from Data Set 10 in Appendix B). Construct a 90% confidence interval estimate of the difference between the mean tar content of unfiltered king size cigarettes and the mean tar content of filtered 10mm cigarettes. Does the result suggest that 100 mm filtered cigarettes have less tar than unfiltered king size cigarettes?

____

____

9-12. ***Hypothesis Test for Cigarette Tar*** Refer to the sample data in Experiment 9-11 and use a 0.05 significance level to test the claim that unfiltered king size cigarettes have a mean tar content greater than that of filtered 100 mm cigarettes. What does the result suggest about the effectiveness of cigarette filters?

Test statistic: __________ *P*–value: __________

Conclusion in your own words: ____

____

9-13. ***BMI for Miss America*** The trend of thinner Miss America winners has generated charges that the contest encourages unhealthy diet habits among young women. Listed below are body mass indexes (BMI) for Miss America winners from two different time periods. Consider the listed values to be simple random samples selected from larger populations.

BMI (from recent winners): 19.5 20.3 19.6 20.2 17.8 17.9 19.1 18.8 17.6 16.8

BMI (from the 1920's and 1930's): 20.4 21.9 22.1 22.3 20.3 18.8 18.9 19.4 18.4 19.1

a. Use a 0.05 significance level to test the claim that recent winners have a lower mean BMI than winners from the 1920's and 1930's.

Test statistic: __________ *P*–value: __________

Conclusion in your own words: ______________________________________

____

b. Construct a 90% confidence interval for the difference between the mean BMI of recent winners and the mean BMI of winners from the 1920's and 1930's.

____

____

9-14. ***Radiation in Baby Teeth*** Listed below are amounts of Strontium-90 (in millibecquerels or mBq per gram of calcium) in a simple random sample of baby teeth obtained from Pennsylvania residents and New York residents born after 1979 (based on data from "An Unexpected Rise in Strontium-90 in U.S. Deciduous Teeth in the 1990s," by Mangano, et. al., *Science of the Total Environment*).

Pennsylvania:	155	142	149	130	151	163	151	142	156	133	138	161
New York:	133	140	142	131	134	129	128	140	140	140	137	143

a. Use a 0.05 significance level to test the claim that the mean amount of Strontium-90 from Pennsylvania residents is greater than the mean amount from New York residents.

Test statistic: __________ *P*–value: __________

Conclusion in your own words: ____

____

b. Construct a 90% confidence interval of the difference between the mean amount of Strontium-90 from Pennsylvania residents and the mean amount from New York residents.

____

____

9-15. ***Longevity*** Listed below are the numbers of years that popes and British monarchs (since 1690) lived after their election or coronation (based on data from *Computer-Interactive Data Analysis*, by Lunn and McNeil, John Wiley & Sons). Treat the values as simple random samples from a larger population.

Popes: 2 9 21 3 6 10 18 11 6 25 23 6 2 15 32 25
11 8 17 19 5 15 0 26

Kings and Queens: 17 6 13 12 13 33 59 10 7 63 9 25 36 15

a. Use a 0.01 significance level to test the claim that the mean longevity for popes is less the mean for British monarchs after coronation.

Test statistic: __________ *P*–value: __________

Conclusion in your own words: ____

____

(*continued*)

b. Construct a 98% confidence interval of the difference between the mean longevity of popes and the mean longevity for kings and queens. What does the result suggest about those two means?

____

____

9-16. ***Sex and Blood Cell Counts*** White blood cell counts are helpful for assessing liver disease, radiation, bone marrow failure, and infectious diseases. Listed below are white blood cell counts found in simple random samples of males and females (based on data from the Third National Health and Nutrition Examination Survey).

Female: 8.90 6.50 9.45 7.65 6.40 5.15 16.60 5.75 11.60 5.90 9.30 8.55 10.80 4.85 4.90 8.75 6.90 9.75 4.05 9.05 5.05 6.40 4.05 7.60 4.95 3.00 9.10

Male: 5.25 5.95 10.05 5.45 5.30 5.55 6.85 6.65 6.30 6.40 7.85 7.70 5.30 6.50 4.55 7.10 8.00 4.70 4.40 4.90 10.75 11.00 9.60

a. Use a 0.01 significance level to test the claim that females and males have different mean white blood cell counts.

Test statistic: __________ *P*–value: __________

Conclusion in your own words: ______________________________

____

b. Construct a 98% confidence interval of the difference between the mean white blood cell count of females and males. Based on the result, does there appear to be a difference?

____

____

9-17. ***Baseline Characteristics*** Reports of results from clinical trials often include statistics about "baseline characteristics," so that we can see that different groups have the same basic characteristics. Refer to Data Set 1 in Appendix B and construct a 95% confidence interval estimate of the difference between the mean age of men and the mean age of women. Based on the result, does it appear that the sample of men and the sample of women are from populations with the same mean?

Test statistic: __________ *P*–value: __________

Conclusion in your own words: ______________________________

____

(continued)

b. Construct a 95% confidence interval estimate of the difference between the mean age of men and the mean age of women. What does the confidence interval suggest?

____

____

9-18 ***Appendix B Data Set: Word Counts*** Refer to Data Set 17 in Appendix B. Use the word counts for male and female psychology students recruited in Mexico (see the columns labeled 3M and 3F).

a. Use a 0.05 significance level to test the claim that male and female psychology students speak the same mean number of words in a day.

Test statistic: __________ *P*–value: __________

Conclusion in your own words: ______________________________

____

b. Construct a 95% confidence interval estimate of the difference between the mean number of words spoken in a day by male and female psychology students in Mexico. Do the confidence interval limits include 0, and what does that suggest about the two means?

____

____

9-19. ***Appendix B Data Set: Voltage*** Refer to Data Set 18 in Appendix B. Use a 0.05 significance level to test the claim that the sample of home voltages and the sample of generator voltages are from populations with the same mean. If there is a statistically significant difference, does that difference have practical significance?

Test statistic: __________ *P*–value: __________

Conclusion in your own words: ______________________________

____

9-20. ***Appendix B Data Set: Weights of Coke*** Refer to Data Set 19 in Appendix B and test the claim that because they contain the same amount of cola, the mean weight of cola in cans of regular Coke is the same as the mean weight of cola in cans of Diet Coke. If there is a difference in the mean weights, identify the most likely explanation for that difference.

Test statistic: __________ *P*–value: __________

Conclusion in your own words: ____

____

9-21. ***Does BMI Change During Freshman Year?*** Listed below are body mass indices (BMI) of students. The BMI of each student was measured in September and April of the freshman year (based on data from "Changes in Body Weight and Fat Mass of Men and Women in the First Year of College: A Study of the 'Freshman 15'," by Hoffman, Policastro, Quick, and Lee, *Journal of American College Health*, Vol. 55, No. 1). Use a 0.05 significance level to test the claim that the mean change in BMI for all students is equal to 0. Does BMI appear to change during freshman year?

April BMI	20.15	19.24	20.77	23.85	21.32
September BMI	20.68	19.48	19.59	24.57	20.96

Test statistic: __________ *P*–value: __________

Conclusion in your own words: ____

____

9-22. ***Confidence Interval for BMI Changes*** Use the same paired data from Experiment 9-21 to construct a 95% confidence interval estimate of the change in BMI during freshman year. Does the confidence interval include 0, and what does that suggest about BMI during freshman year?

____

____

9-23. ***Are Best Actresses Younger than Best Actors?*** Listed below are ages of actresses and actors at the times that they won Oscars. The data are paired according to the years that they won. Use a 0.05 significance level to test the common belief that best actresses are younger than best actors. Does the result suggest a problem in our culture?

Best Actresses	28	32	27	27	26	24	25	29	41	40	27	42	33	21	35
Best Actors	62	41	52	41	34	40	56	41	39	49	48	56	42	62	29

Test statistic: __________ *P*–value: __________

Conclusion in your own words: ____

____

9-24. ***Are Flights Cheaper When Scheduled Earlier?*** Listed below are the costs (in dollars) of flights from New York (JFK) to San Francisco for US Air, Continental, Delta, United, American, Alaska, and Northwest. Use a 0.01 significance level to test the claim that flights scheduled one day in advance cost more than flights scheduled 30 days in advance. What strategy appears to be effective in saving money when flying?

Flight scheduled one day in advance	456	614	628	1088	943	567	536
Flight scheduled 30 days in advance	244	260	264	264	278	318	280

Test statistic: __________ *P*–value: __________

Conclusion in your own words: ____

____

9-25. ***Does Your Body Temperature Change During the Day?*** Listed below are body temperatures (in ^{o}F) of subjects measured at 8 AM and at 12 AM (from University of Maryland physicians listed in Data Set 2 in Appendix B). Construct a 95% confidence interval estimate of the difference between the 8 AM temperatures and the 12 AM temperatures. Is body temperature basically the same at both times?

8 AM	97.0	96.2	97.6	96.4	97.8	99.2
12 AM	98.0	98.6	98.8	98.0	98.6	97.6

Test statistic: __________ *P*–value: __________

Conclusion in your own words: ____

____

9-26. ***Is Blood Pressure the Same for Both Arms?*** Listed below are systolic blood pressure measurements (mm Hg) taken from the right and left arms of the same woman (based on data from "Consistency of Blood Pressure Differences Between the Left and Right Arms," by Eguchi et al, *Archives of Internal Medicine*, Vol. 167). Use a 0.05 significance level to test for a difference between the measurements from the two arms. What do you conclude?

Right arm	102	101	94	79	79
Left arm	175	169	182	146	144

Test statistic: __________ *P*–value: __________

Conclusion in your own words: ____

____

9-27. ***Appendix B Data Set: Paper or Plastic?*** Refer to Data Set 23 in Appendix B. Construct a 95% confidence interval estimate of the mean of the differences between weights of discarded paper and weights of discarded plastic. Which seems to weigh more: discarded paper or discarded plastic?

____

____

9-28. ***Appendix B Data Set: Glass and Food*** Refer to Data Set 23 in Appendix B. Construct a 95% confidence interval estimate of the mean of the differences between weights of discarded glass and weights of discarded food. Which seems to weigh more: discarded glass or discarded food?

____

____

9-29. ***Testing Effects of Alcohol*** Researchers conducted an experiment to test the effects of alcohol. The errors were recorded in a test of visual and motor skills for a treatment group of 22 people who drank ethanol and another group of 22 people given a placebo. The errors for the treatment group have a standard deviation of 2.20, and the errors for the placebo group have a standard deviation of 0.72 (based on data from "Effects of Alcohol Intoxication on Risk Taking, Strategy, and Error Rate in Visuomotor Performance," by Streufert, et al., *Journal of Applied Psychology,* Vol. 77, No. 4). Use a 0.05 significance level to test the claim that the treatment group has errors that vary more than the errors of the placebo group.

Test statistic: __________ *P*–value: __________

Conclusion in your own words: ____

____

9-30. ***Heights*** Listed below are heights (cm) of randomly selected females and males taken from Data Set 1 in Appendix B. Use a 0.05 significance level to test the claim that females and males have heights with the same amount of variation.

Female	163.7	165.5	163.1	166.3	163.6	170.9	153.5	155.7	153.0	157.0
Male	178.8	177.5	187.8	172.4	181.7	169.0	186.9	183.1	176.4	183.4

Test statistic: __________ *P*–value: __________

Conclusion in your own words: ____

____

9-31. ***Freshman 15 Study*** Use the sample weights (kg) of male and female college students measured in April of their freshman year, as listed in Data Set 4 in Appendix B. Use a 0.05 significance level to test the claim that near the end of the freshman year, weights of male college students vary more than weights of female college students.

Test statistic: __________ *P*–value: __________

Conclusion in your own words: ____

____

9-32. ***M&Ms*** Refer to Data Set 20 in Appendix B and use the weights (g) of the red M&Ms and the orange M&Ms. Use a 0.05 significance level to test the claim that the two samples are from populations with the same amount of variation.

Test statistic: __________ *P*–value: __________

Conclusion in your own words: ____

____

10

Correlation and Regression

10-1 Minitab's Assistant Feature

Minitab introduced its new *Assistant* feature in Release 16. If you click on the main menu item of **Assistant,** you get a submenu that includes the option of **Regression.** If you click on that item, you will get a screen that includes the following.

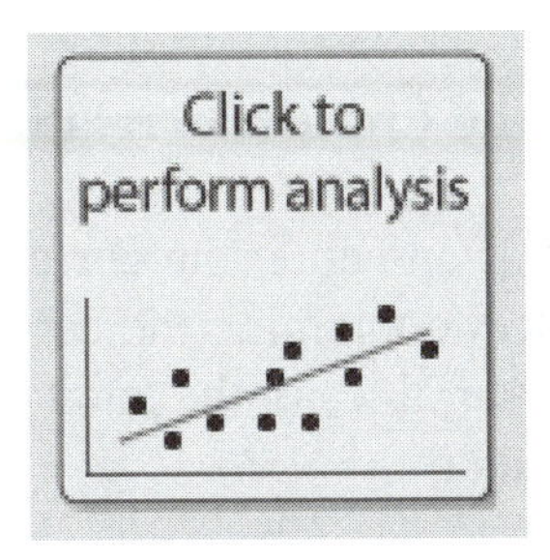

Clicking on the above box starts a process for an in-depth analysis of correlation and regression. You must enter the columns containing the lists of data to be used for the x and y variables, and results will be displayed in four different screens providing a wealth of information, including the linear correlation coefficient, P-value, scatterplot, equation of the regression line, a "report card" with comments about normality, unusual points, sample size, a diagnostic report that uses residuals, and a "model selection" report. The following sections of this manual/workbook describe the use of Minitab's basic functions, but consider using the *Assistant* feature in place of those basic methods.

10-2 Scatterplot

Chapter 10 in the Triola textbook introduces the basic concepts of linear correlation and regression. The basic objective is to use paired sample data to determine whether there is a relationship between two variables and, if so, identify what the relationship is. Consider the paired sample data in the table below. The data consist of measured shoe print lengths and heights of five males. We want to determine whether there is a relationship between shoe print length and height. If such a relationship exists, we want to identify it with an equation so that we can predict the height of a male based on the length of his shoe print.

Table 10-1 Shoe Print Lengths and Heights of Males

Shoe Print (cm)	29.7	29.7	31.4	31.8	27.6
Height (cm)	175.3	177.8	185.4	175.3	172.7

Before using concepts of correlation and regression, it is always wise to construct a scatterplot so that we can see if there appears to be a relationship between the two variables, and we can see if the pattern of points is approximately the pattern of a straight line. The Minitab procedure for constructing a scatterplot was given in Section 2-4 of this manual/workbook. Enter the paired data in columns of Minitab or open a Minitab worksheet, select **Graph**, then **Scatterplot**. Select the option of **Simple**, then click **OK**.

10-3 Correlation

The textbook describes the linear correlation coefficient r as a measure of the strength of the linear relationship between two variables. Minitab can compute the value of r for paired data. Instead of performing complicated manual calculations, follow these steps to let Minitab calculate r.

Minitab Procedure for Calculating the Linear Correlation Coefficient r

1. Enter the x values in column C1, and enter the corresponding y values in column C2 (or open a Minitab worksheet containing the columns of paired data). It is also wise to enter the names of the variables in the empty cells located immediately above the first row of data values.

2. Select **Stat**, then **Basic Statistics**, then **Correlation**.

3. In the dialog box, enter the column labels of C1 C2 (or enter the variable names), then click **OK**.

Shown below is the Minitab display that results from using the data in the above table. The variable names of ShoePrint and Height were also entered. Because the P-value of 0.294 is greater than the significance level of 0.05, we conclude that there is not sufficient evidence to support the claim of a linear correlation between the lengths of shoe prints and heights of males.

Correlations: ShoePrint, Height

```
Pearson correlation of ShoePrint and Height = 0.591
P-Value = 0.294
```

10-4 Regression

The textbook discusses the important topic of regression. Given a collection of paired sample data, the **regression equation**

$$\hat{y} = b_0 + b_1 x$$

describes the relationship between the two variables. The graph of the regression equation is called the **regression line** (or *line of best fit*, or *least-squares line*). The regression equation expresses a relationship between x (called the *independent variable* or *predictor variable*) and y (called the *dependent variable* or *response variable*). The y-intercept of the line is b_0 and the slope is b_1. Given a collection of paired data, we can find the regression equation as follows.

Minitab Procedure for Finding the Equation of the Regression Line

1. Enter the x values in column C1, and enter the corresponding y values in column C2 (or open a Minitab worksheet containing the columns of paired data). It is also wise to enter the names of the variables in the empty cells located immediately above the first row of data values.

2. Select **Stat**, then **Regression**, then **Regression** once again.

3. Make these entries in the dialog box:

    - Enter C2 (the y values) in the box for the "Response" variable.
    - Enter C1 (the x values) in the box for the "Predictor" variable.
    - Click on the **Options** bar and click on the "**Fit intercept**" box.
    - Click on the **Results** bar and select the second option that includes the regression equation, s, and R-Squared.
    - Click **OK**.

Using the sample data in the preceding table, if you use the response variable of Height and the predictor variable of ShoePrint, the Minitab display will be as follows.

Regression Analysis: Height versus ShoePrint

```
The regression equation is
Height = 125 + 1.73 ShoePrint


Predictor      Coef  SE Coef     T      P
Constant     125.41    40.92  3.07  0.055
ShoePrint     1.727    1.360  1.27  0.294


S = 4.53876   R-Sq = 35.0%   R-Sq(adj) = 13.3%


Analysis of Variance

Source          DF     SS     MS     F      P
Regression       1  33.22  33.22  1.61  0.294
Residual Error   3  61.80  20.60
Total            4  95.02
```

This display includes the regression equation, which we express as $\hat{y} = 125 + 1.73x$. The Minitab display also includes "R-sq = 35.0%," which is the coefficient of determination defined in the textbook; that value indicates that 35.0% of the variation in heights can be explained by the shoe print lengths, and 65.0% of the variation in heights remains unexplained. The display also includes other items discussed later in the textbook.

In addition to the *equation* of the regression line, we can also use the following procedure to get a *graph* of the regression line superimposed on the scatter diagram.

Minitab Procedure for Graph of Regression Line

1. Enter the x values in column C1, and enter the corresponding y values in column C2 (or open a Minitab worksheet containing the columns of paired data). It is also wise to enter the names of the variables in the empty cells located immediately above the first row of data values.
2. Select **Stat**, **Regression**, then **Fitted line plot**.
3. Complete the dialog box by entering the response (y) variable, the predictor (x) variable, and select the **Linear** model. Click **OK**.

Shown below is the graph of the regression line for the data in the preceding table. We can see that the regression line doesn't fit the points very well.

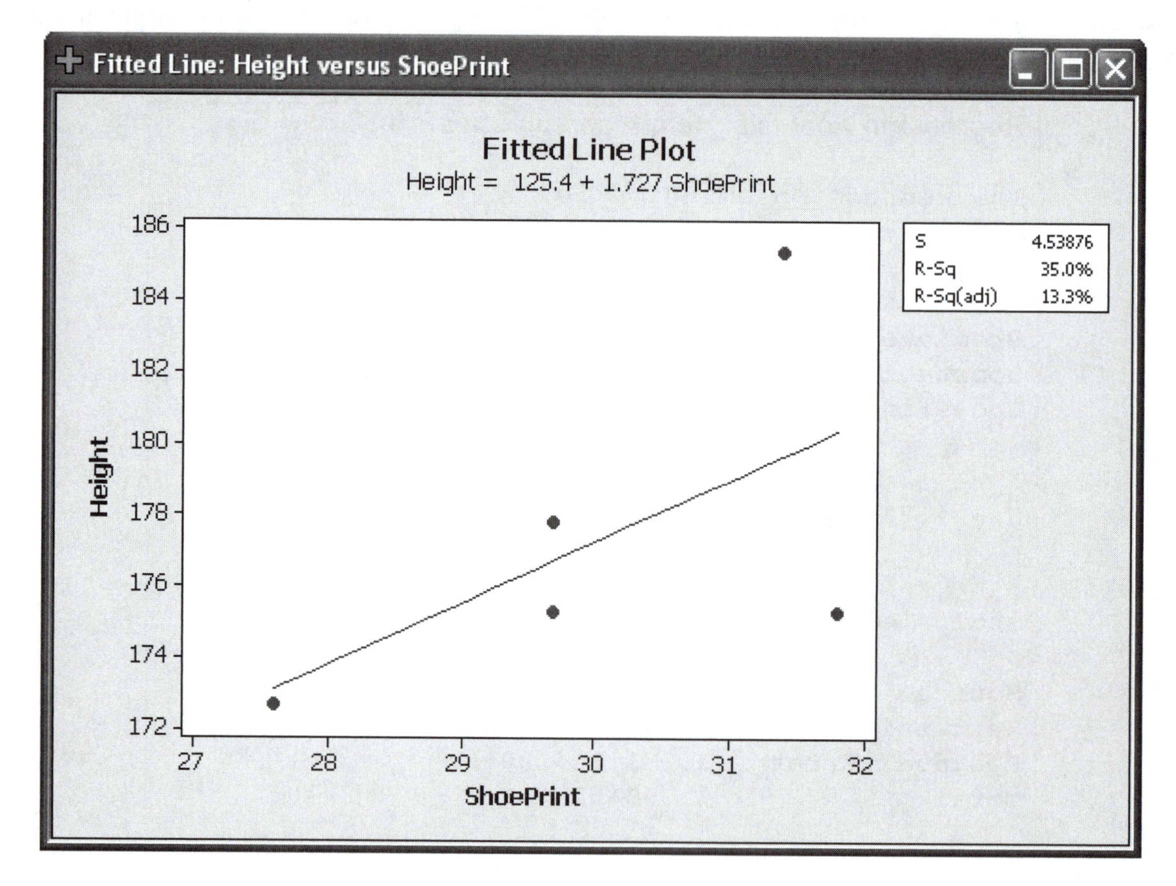

10-5 Predictions

Important: When making predictions, do not blindly substitute values into the regression equation. See Section 10-3 in the textbook, and see Figure 10-5 reproduced here.

Is the regression equation a good model?

- **The regression line graphed in the scatterplot shows that the line fits the points well.**
- ***r* indicates that there is a linear correlation**
- **The prediction is not much beyond the scope of the available sample data.**

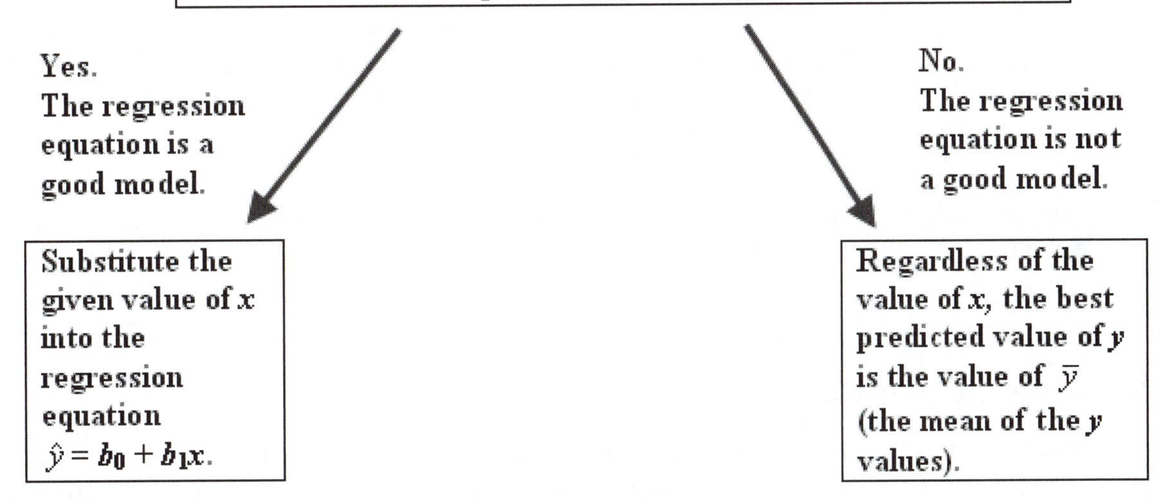

We can use Minitab for making predictions by substituting a value of x into the regression equation. Simply follow the same procedure for finding the equation of the regression line as given in Section 10-4 of this manual/wokbook (using **Stat/Regression/Regression**) with this addition:

In the **Options** box:

1. Enter the x value(s) in the box labeled "Prediction intervals for new observations."
2. Enter the desired confidence level, such as 95 (for 95% confidence).

In part (b) of Example 4 in the textbook, 40 pairs of sample data are used to predict height given a shoe print length of 29 cm. Here is the Minitab result:

```
Predicted Values for New Observations

New Obs      Fit   SE Fit          95% CI                   95% PI
      1  174.269    0.940   (172.366, 176.171)   (162.087, 186.451)


Values of Predictors for New Observations

          Shoe
New Obs  Print
      1   29.0
```

The above results show that if we substitute $x = 29$ into the regression equation, the best predicted value of y is 174.269 cm (or 174.3 cm when rounded). A 95% prediction interval is

$$162.1 \text{ cm} < y < 186.5 \text{ cm}$$

The values shown in the display for "95% C.I." are confidence interval limits for the mean height for males who are 29 cm tall.

10-6 Multiple Regression

The textbook discusses multiple regression, and Minitab can provide multiple regression results. Once a collection of sample data has been entered, you can easily experiment with different combinations of columns (variables) to find the combination that is best. Here is the Minitab procedure.

Minitab Procedure for Multiple Regression

1. Enter the sample data in columns C1, C2, C3, . . . (or open a Minitab worksheet containing the columns of sample data). Also enter the names of the variables in the empty cells located immediately above the first row of sample values.

2. Click on **Stat**, **Regression**, then **Regression** once again.

3. Make these entries in the dialog box:
    - Enter the column containing the y values in the box for the "Response" variable.
    - In the box for the "Predictor" variables, enter the columns that you want included for the independent variables (the x variables).
    - Click on the **Options** bar and click on the "**Fit intercept**" box.
    - Click on the **Results** bar and select the second option that includes the regression equation, s, and R-Squared, then click **OK**.

As an example, the table below includes measurements obtained from anesthetized bears. If you enter the sample data included in the table and select column C1 as the response variable and columns C3 and C6 as predictor variables, the Minitab results will be as shown below the table. The results correspond to this multiple regression equation:

$$\hat{y} = -374 + 18.8x_3 + 5.87x_6$$

The results also include the adjusted coefficient of determination (75.9%) as well as a *P*-value (0.012) for overall significance of the multiple regression equation.

Data from Anesthetized Male Bears

Var.	Minitab Column	Name	Sample Data							
y	C1	WEIGHT	80	344	416	348	262	360	332	34
x_2	C2	AGE	19	55	81	115	56	51	68	8
x_3	C3	HEADLEN	11.0	16.5	15.5	17.0	15.0	13.5	16.0	9.0
x_4	C4	HEADWDTH	5.5	9.0	8.0	10.0	7.5	8.0	9.0	4.5
x_5	C5	NECK	16.0	28.0	31.0	31.5	26.5	27.0	29.0	13.0
x_6	C6	LENGTH	53.0	67.5	72.0	72.0	73.5	68.5	73.0	37.0
x_7	C7	CHEST	26	45	54	49	41	49	44	19

Here is the Minitab screen showing the entry of the values in the above table:

↓	C1 WEIGHT	C2 AGE	C3 HEADLEN	C4 HEADWTH	C5 NECK	C6 LENGTH	C7 CHEST
1	80	19	11.0	5.5	16.0	53.0	26
2	344	55	16.5	9.0	28.0	67.5	45
3	416	81	15.5	8.0	31.0	72.0	54
4	348	115	17.0	10.0	31.5	72.0	49
5	262	56	15.0	7.5	26.5	73.5	41
6	360	51	13.5	8.0	27.0	68.5	49
7	332	68	16.0	9.0	29.0	73.0	44
8	34	8	9.0	4.5	13.0	37.0	19

```
The regression equation is
WEIGHT = - 374 + 18.8 HEADLEN + 5.87 LENGTH

Predictor          Coef      SE Coef         T        P
Constant         -374.3        134.1     -2.79    0.038
HEADLEN           18.82        23.15      0.81    0.453
LENGTH            5.875        5.065      1.16    0.299

S = 68.5649      R-Sq = 82.8%      R-Sq(adj) = 75.9%

Analysis of Variance

Source            DF         SS          MS          F          P
Regression         2     113142       56571      12.03      0.012
Residual Error     5      23506        4701
Total              7     136648
```

We can again use Minitab to predict *y* values for given values of the predictor variables. For example, to predict a bear's weight (*y* value), given that its head length (HEADLEN) is 14 in. and its LENGTH is 71.0 in., follow the same steps listed earlier in this section, but click on "Options" before clicking on the **OK** box. In the second dialog box, enter the values of 14.0 and 71.0 in the box identified as "Prediction intervals for new observations," then enter the desired confidence level, such as 95 (for 95% confidence). In addition to the elements in the above Minitab display, the following will also be shown:

```
Predicted Values for New Observations

New Obs     Fit    SE Fit        95.0% CI              95.0% PI
1         306.3      43.9   (  193.4,    419.2)   (   97.0,    515.6)
```

This portion of the display shows that for a bear with a head length of 14.0 in. and a length of 71.0 in., the predicted weight is 306.3 lb. The prediction interval (for a single weight) is given, and the confidence interval (for the mean weight of all bears with 14.0 in. head length and 71.0 in. length) is also given.

When trying to find the best multiple regression equation, Minitab's **Stepwise Regression** procedure may be helpful. Select **Stat**, **Regression, Stepwise Regression**.

10-7 Nonlinear Regression

Nonlinear regression is discussed in the Triola statistics textbooks (except *Essentials of Statistics*). The objective is to find a mathematical function that "fits" or describes real-world data. Among the models discussed in the textbook, we will describe how Minitab can be used for the linear, quadratic, logarithmic, exponential, and power models.

To illustrate the use of Minitab, consider the sample data in the table below. As in the textbook, we will use the coded year values for x, so that $x = 1, 2, 3, \ldots, 11$. The y values are the populations (in millions) of 5, 10, 17, ..., 281. For the following models, we enter the coded years 1, 2, 3, . . . , 11 in column C1 and we enter 5, 10, 17, . . . , 281 in column C2. The resulting models are based on the coded values of x (1, 2, 3, . . . , 11) and the population values in millions (5, 10, 17, . . . , 281).

Population (in millions) of the United States

Year	1800	1820	1840	1860	1880	1900	1920	1940	1960	1980	2000
Coded year	1	2	3	4	5	6	7	8	9	10	11
Population	5	10	17	31	50	76	106	132	179	227	281

Linear Model: $y = a + bx$

The linear model can be obtained by using Minitab's correlation and regression module. The procedure is described in Section 10-4 of this manual/workbook. For the data in the above table, enter the coded year values of 1, 2, 3, . . . , 11 in column C1 and enter the population values of 5, 10, 17, . . . , 281 in column C2. The result will be $y = -61.9 + 2.72x$ with $R^2 = 92.5\%$ (or 0.925). The high value of R^2 suggests that the linear model is a reasonably good fit.

Quadratic Model: $y = ax^2 + bx + c$

With the coded values of x in column C1 and the population values (in millions) in column C2, select **Stat**, **Regression**, **Fitted Line Plot**. In the dialog box, enter C2 for the response variable, enter C1 for the predictor variable, and click on the **Quadratic** option. The results show that the function has the form given as $y = 10.01 - 6.003x + 2.767x^2$ with $R^2 = 0.999$. This higher value of R^2 suggests that the quadratic model is a better fit than the linear model.

Logarithmic Model: $y = a + b \ln x$

Minitab does not have a direct procedure for finding a logarithmic model, but there are ways to get it. One approach is to replace the coded x values (1, 2, 3, . . . , 11) with $\ln x$ values (ln 1, ln 2, ln 3, . . . , ln 11) This can be accomplished by using the command editor in the session window; enter the command LET C1 = LOGE(C1). [You could also use **Calc/Calculator** with the expression LOGE(C1).] After replacing the x values with $\ln x$ values, use **Stat**, **Regression**, **Fitted Line Plot** as in Section 10-4. That is, use the same procedure for finding the linear regression equation. The result is $y = -65.89 + 105.1 \ln x$, with $R^2 = 0.696$, suggesting that this model does not fit as well as the linear or quadratic models. Of the three models considered so far, the quadratic model appears to be best.

Exponential Model: $y = ab^x$

The exponential model is tricky, but it can be obtained using Minitab. Enter the values of x in the first column, and enter the values of ln y in the second column. The values of ln y can be found by using the command editor in the session window; enter the command LET C2 = LOGE(C2). [You could also use **Calc/Calculator** with the expression LOGE(C2).]

After replacing the y values with ln y values, use **Stat**, **Regression**, **Regression** to find the linear regression equation as described in Section 10-4 of this book. When you get the results from Minitab, the value of the coefficient of determination is correct, but the values of a and b in the exponential model must be computed as follows:

To find the value of a: Evaluate e^{b_0} where b_0 is the y-intercept given by Minitab.

To find the value of b: Evaluate e^{b_1} where b_1 is the slope given by Minitab.

Using the data in the above table, the value of $R^2 = 0.963$ is OK as is, but Minitab's regression equation of $y = 1.6556 + 0.39405x$ must be converted as follows.

$$a = e^{b_0} = e^{1.6556} = 5.2362$$
$$b = e^{b_1} = e^{0.39405} = 1.4830$$

Using these values of a and b, we express the exponential model as $y = 5.2362(1.4830^x)$.

Power Model: $y = ax^b$

Replace the x values with ln x values and replace the y values with ln y values. The values of ln x can be found by using the command editor in the session window; enter the command LET C1 = LOGE(C1). [You could also use **Calc/Calculator** with the expression LOGE(C1).] The values of ln y can be found the same way. Then use **Stat**, **Regression**, **Regression** to find the linear regression equation as described in Section 10-4 of this workbook. When you get the results from Minitab, the value of the coefficient of determination is correct, the value of b is the same as the slope of the regression line, but the y-intercept must be converted as follows.

To find the value of a: Evaluate e^{b_0} where b_0 is the y-intercept given by Minitab.

Using the data in the above table, we get $R^2 = 0.976$. From Minitab's linear regression equation of $y = 1.2099 + 1.76606x$, we get $b = 1.76606$ and a is computed as follows.

$$a = e^{b_0} = e^{1.2099} = 3.35315$$

Using these values of a and b, we express the power model as $y = 3.35315(x^{1.76606})$.

The rationale underlying the methods for the exponential and power models is based on transformations of equations. In the exponential model of $y = ab^x$, for example, taking natural logarithms of both sides yields ln y = ln a + x (ln b), which is the equation of a straight line. Minitab can be used to find the equation of this straight line that fits the data best; the intercept will be ln a and the slope will be ln b, but we need the values of a and b, so we solve for them as described above. Similar reasoning is used with the power model.

CHAPTER 10 EXPERIMENTS: Correlation and Regression

10–1. ***Bear Weights and Ages*** Refer to the Bears data set in Appendix B of the textbook (Data Set 7). (The Minitab worksheet is BEARS.) Use the values for WEIGHT (x) and the values for AGE (y) to find the following.

a. Display the scatter diagram of the paired WEIGHT/AGE data. Based on that scatter diagram, does there appear to be a relationship between the weights of bears and their ages? If so, what is it?

____

b. Find the value of the linear correlation coefficient r. __________

c. Assuming a 0.05 level of significance, what do you conclude about the correlation between weights and ages of bears?

____

d. Find the equation of the regression line. (Use WEIGHT as the x predictor variable, and use AGE as the y response variable.) ____________________

e. What is the best predicted age of a bear that weighs 300 lb?________________

10–2. ***Effect of Transforming Data*** The ages used in Experiment 10–1 are in months. Convert them to days by multiplying each age by 30, then repeat Experiment 10–1 and enter the responses here:

a. Display the scatter diagram of the paired WEIGHT/AGE data. Based on that scatter diagram, does there appear to be a relationship between the weights of bears and their ages? If so, what is it?

____

b. Find the value of the linear correlation coefficient r.________________

c. Assuming a 0.05 level of significance, what do you conclude about the correlation between weights and ages of bears?

____

d. Find the equation of the regression line. (Use WEIGHT as the x predictor variable, and use AGE as the y response variable.) ____________________

e. What is the best predicted age of a bear that weighs 300 lb?________________

f. After comparing the responses obtained in Experiment 10-1 to those obtained here, describe the general effect of changing the scale for one of the variables.

____

____

10–3. ***Bear Weights and Chest Sizes*** Refer to the Bears data set in Appendix B of the textbook. (The Minitab worksheet is BEARS.) Use the values for CHEST (x) and the values for WEIGHT (y) to find the following.

a. Display the scatter diagram of the paired CHEST/WEIGHT data. Based on that scatter diagram, does there appear to be a relationship between the chest sizes of bears and their weights? If so, what is it?

____

b. Find the value of the linear correlation coefficient r. __________

c. Assuming a 0.05 level of significance, what do you conclude about the correlation between chest sizes and weights of bears?

____

d. Find the equation of the regression line. (Use CHEST as the x predictor variable, and use WEIGHT as the y response variable.) ____________________

e. What is the best predicted weight of a bear with a chest size of 36.0 in?________

f. When trying to obtain measurements from an anesthetized bear, what is a practical advantage of being able to predict the bear's weight by using its chest size?

____

____

10–4. ***Effect of No Variation for a Variable*** Use the following paired data and obtain the indicated results.

x	1	2	3	4	5	7	7	9
y	5	5	5	5	5	5	5	5

a. Print a scatter diagram of the paired x and y data. Based on the result, does there appear to be a relationship between x and y? If so, what is it?

____

b. What happens when you try to find the value of r? Why?

____

c. What do you conclude about the correlation between x and y? What is the equation of the regression line?

____

10–5. ***Town Courts*** Listed below are amounts of court income and salaries paid to the town justices (based on data from the *Poughkeepsie Journal*). All amounts are in thousands of dollars, and all of the towns are in Dutchess County, New York.

Court Income	65	404	1567	1131	272	252	111	154	32
Justice Salary	30	44	92	56	46	61	25	26	18

Is there a correlation between court incomes and justice salaries? Explain.

____

What is the equation of the regression line? ______________________

Find the best predicted justice salary for a court with income of $83,941 (or $83.941 thousand). ____________

10–6. ***Car Repair Costs*** Listed below are repair costs (in dollars) for cars crashed at 6 mi/h in full-front crash tests and the same cars crashed at 6 mi/h in full-rear crash tests (based on data from the Insurance Institute for Highway Safety). The cars are the Toyota Camry, Mazda 6, Volvo S40, Saturn Aura, Subaru Legacy, Hyundai Sonata, and Honda Accord.

Front	936	978	2252	1032	3911	4312	3469
Rear	1480	1202	802	3191	1122	739	2767

Is there a correlation between front and rear repair costs? Explain.

____

What is the equation of the regression line? ______________________

Find the best predicted repair cost for a full-rear crash for a Volkwagon Passat, given that its repair cost from a full-front crash is $4594. ____________

10–7. ***Movie Data*** Refer to the data from movies in the table below, where all amounts are in millions of dollars. Let the movie gross amount be the dependent y variable.

Budget	62	90	50	35	200	100	90
Gross	65	64	48	57	601	146	47

Is there a correlation between the budget amount and the gross amount? Explain.

____

What is the equation of the regression line? ______________________

Find the best predicted gross amount for a movie with a budget of $150 million.______

10–8. ***Crickets and Temperature*** The numbers of chirps in one minute were recorded for different crickets, and the corresponding temperatures were also recorded. The results are given in the table below.

Chirps in one minute	882	1188	1104	864	1200	1032	960	900
Temperature (°F)	69.7	93.3	84.3	76.3	88.6	82.6	71.6	79.6

Is there a correlation between the number of chirps and the temperature? Explain.

____

What is the equation of the regression line? ____________________________

Find the best predicted temperature when a cricket chirps 1000 times in one minute.___

10–9. ***Blood Pressure*** Refer to the systolic and diastolic blood pressure measurements from randomly selected subjects. Let the dependent variable y represent diastolic blood pressure.

Systolic	138	130	135	140	120	125	120	130	130	144	143	140	130	150
Diastolic	82	91	100	100	80	90	80	80	80	98	105	85	70	100

Is there a correlation between systolic and diastolic blood pressure? Explain.

____

What is the equation of the regression line? ____________________________

Find the best predicted measurement of diastolic blood pressure for a person with a systolic blood pressure of 123. ________________

10–10. ***Garbage Data for Predicting Household Size*** The "Garbage" data set (Data Set 23) in Appendix B of the textbook consists of data from the Garbage Project at the University of Arizona. (The Minitab worksheet is GARBAGE.) Use household size (HHSIZE) as the response y variable. For each predictor x variable given below, find the value of the linear correlation coefficient, the equation of the regression line, and the value of the coefficient of determination r^2. Enter the results in the spaces below.

(continued)

	r	Equation of regression line	r^2
Metal	___	______________________	___
Paper	___	______________________	___
Plastic	___	______________________	___
Glass	___	______________________	___
Food	___	______________________	___
Yard	___	______________________	___
Text	___	______________________	___
Other	___	______________________	___
Total	___	______________________	___

Based on the above results, which single independent variable appears to be the best predictor of household size? Why?

____

____

10–11. Use the same data set described in Experiment 10-10. Let household size (HHSIZE) be the dependent y variable and use the given predictor x variables to fill in the results below.

	Multiple regression eq.	R^2	Adj. R^2
Metal and Paper	____________________	___	_____
Plastic and Food	____________________	___	_____
Metal, Paper, Glass	____________________	___	_____
Metal, Paper, Plastic, Glass	____________________	___	_____

Based on the above results, which of the multiple regression equations appears to best fit the data? Why?

____

____

10-12. ***Appendix B Data Set: Predicting Nicotine in Cigarettes*** Refer to Data Set 4 in Appendix B and use the tar, nicotine, and CO amounts for the cigarettes that are 100 mm long, filtered, non-menthol, and non-light (the last set of measurements). Find the best regression equation for predicting the amount of nicotine in a cigarette. Is the best regression equation a good regression equation for predicting the nicotine content?

10-13. ***Predicting IQ Score*** Refer to Data Set 6 in Appendix B and find the best regression equation with IQ score as the response (y) variable. Use predictor variables of brain volume and/or weight. Why is this equation best? Based on these results, can we predict someone's IQ score if we know the volume and weight of their brain? Based on these results, does it appear that people with larger brains have higher IQ scores?

10-14. ***Full IQ Score*** Refer to Data Set 5 in Appendix B and find the best regression equation with IQF (full IQ score) as the response (y) variable. Use predictor variables of IQV (verbal IQ score) and IQP (performance IQ score). Why is this equation best? Based on these results, can we predict someone's full IQ score if we know their verbal IQ score and their performance IQ score? Is such a prediction likely to be very accurate?

10-15. ***Manatee Deaths from Boats*** Listed below are the numbers of Florida manatee deaths related to encounters with watercraft (based on data from *The New York Times*). The data are listed in order, beginning with the year 1980 and ending with the year 2000.

16 24 20 15 34 33 33 39 43 50 47 53 38 35 49 42 60 54 67 82 78

Use the given data to find equations and coefficients of determination for the indicated models.

	Equation	R^2
Linear	__________	__________
Quadratic	__________	__________
Logarithmic	__________	__________
Exponential	__________	__________
Power	__________	__________

Based on the above results, which model appears to best fit the data? Why?

____

____

What is the best predicted value for 2001? In 2001, there were 82 watercraft–related manatee deaths. How does the predicted value compare to the actual value?

____

____

____

11

Goodness-of-Fit and Contingency Tables

11-1 Goodness-of-Fit

We use a *goodness-of-fit* test with sample data consisting of observed frequency counts arranged in a single row or column (called a one-way frequency table). We test for the claim that the observed frequency counts agree with some claimed distribution, so that there is a *good fit* of the observed data with the claimed distribution.

Minitab Procedure for Goodness-of-Fit

1. First collect the observed values O and enter them in column C1.

2. If the expected frequencies are not all the same, enter the expected *proportions* in column C2.

3. Select **Stat**, **Tables**, and **Chi-Square Goodness-of-Fit Test (One Variable)**..

4. Enter C1 in the "Observed counts" box.

5. Select "Equal proportions" if the expected frequencies are all the same. Otherwise, select "Specific proportions" and enter C2 in the box.

6. Click **OK.**

The textbook includes an example (Example 1 in Section 11-2 of the textbook) with the frequency counts of the 100 last digits of the weights of Californians. Those frequency counts are listed in the table below. We want to test the claim that the observed digits are from a population of weights in which the last digits do not occur with the same frequency.

Last Digits of Weights

Last Digit	Frequency
0	46
1	1
2	2
3	3
4	3
5	30
6	4
7	0
8	8
9	3

After entering the observed frequencies column C1, select **Stat**, **Tables,** select **Chi-Square Goodness-of-Fit Test (One Variable)** and use the dialog box as shown below.

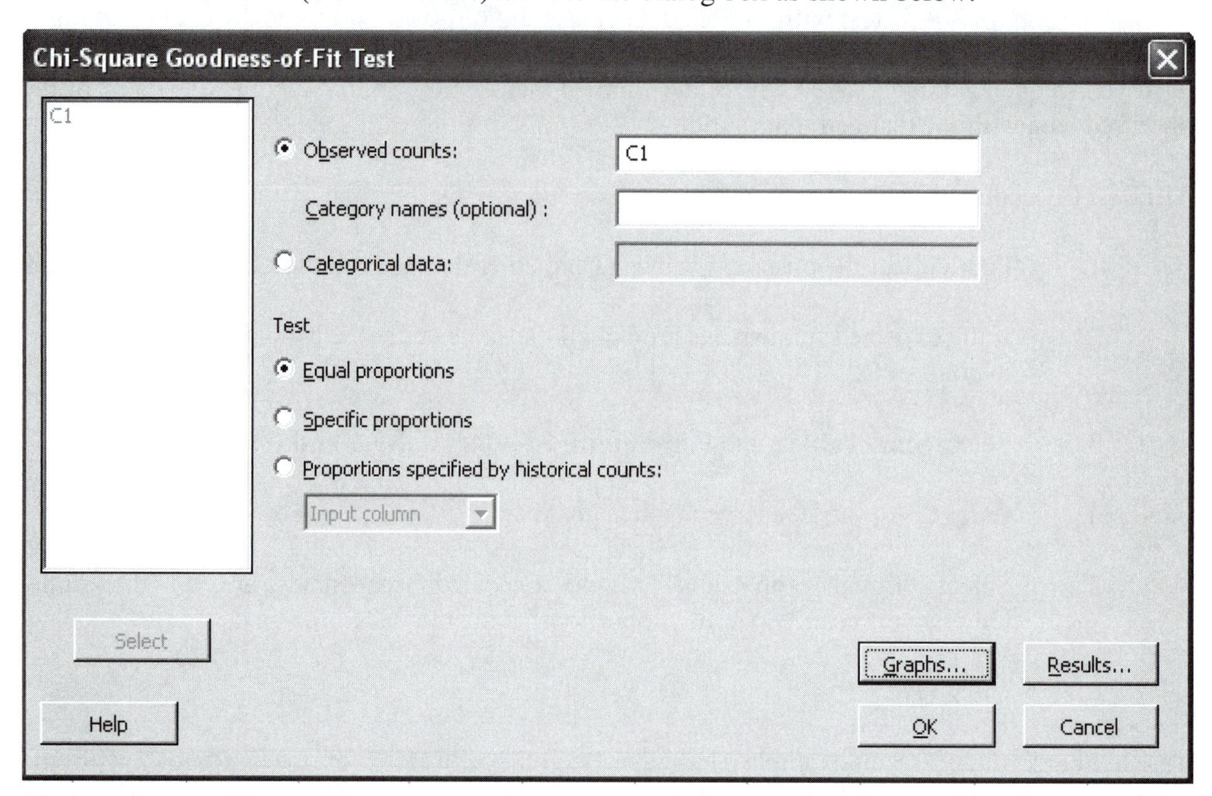

The Minitab results are shown below. The P-value of 0.000 suggests that we reject the null hypothesis that the digits occur with the same frequency. There is sufficient evidence to support the claim that the digits do *not* occur with the same frequency.

Chi-Square Goodness-of-Fit Test for Observed Counts in Variable: C1

Category	Observed	Test Proportion	Expected	Contribution to Chi-Sq
1	46	0.1	10	129.6
2	1	0.1	10	8.1
3	2	0.1	10	6.4
4	3	0.1	10	4.9
5	3	0.1	10	4.9
6	30	0.1	10	40.0
7	4	0.1	10	3.6
8	0	0.1	10	10.0
9	8	0.1	10	0.4
10	3	0.1	10	4.9

N	DF	**Chi-Sq**	**P-Value**
100	9	**212.8**	**0.000**

11-2 Contingency Tables

A **contingency table** (or **two-way frequency table**) is a table in which frequencies correspond to two variables. One variable is used to categorize rows, and a second variable is used to categorize columns. Let's consider the data in the contingency table shown below (from Example 1 in Section 11-3 of the textbook).

Table 11-6 Study of Success with Different Treatments for Stress Fracture

	Success	Failure
Surgery	54	12
Weight-Bearing Cast	41	51
Non-Weight Bearing Cast for 6 Weeks	70	3
Non-Weight Bearing Cast Less than 6 Weeks	17	5

To use Minitab with a contingency table, follow this procedure:

Minitab Procedure for Contingency Tables

1. Enter the individual columns of data in Minitab columns C1, C2, C3, . . . , . The Minitab display shows the data from the above table.

Worksheet 1 ***

↓	C1	C2
1	54	12
2	41	51
3	70	3
4	17	5

2. Select **Stat**, **Tables**, **Chisquare Test (Two-Way Table in Worksheet)**.

3. Proceed to enter the names of the columns containing the observed frequencies. For the above data, enter C1 C2. Click **OK**.

Shown below are the Minitab results from the data in the above table. The test statistic and *P*-value have been highlighted in bold. The *P*-value of 0.000 is less than a 0.05 significance level, so we reject the null hypothesis of independence between the row and column variables. It appears that success or failure is dependent on the treatment.

Chi-Square Test: C1, C2

```
Expected counts are printed below observed counts
Chi-Square contributions are printed below expected counts

           C1      C2  Total
    1      54      12     66
        47.48   18.52
        0.896   2.296

    2      41      51     92
        66.18   25.82
        9.582  24.561

    3      70       3     73
        52.51   20.49
        5.823  14.925

    4      17       5     22
        15.83    6.17
        0.087   0.223

Total     182      71    253

Chi-Sq = 58.393, DF = 3, P-Value = 0.000
```

11-3 Fisher's Exact Test

Fisher's exact test can be used for two-way tables, and it is used mostly for 2 × 2 tables. This test uses an *exact* distribution instead of an approximating chi-square distribution. It is particularly helpful when the approximating chi-square distribution cannot be used because of expected cell frequencies that are less than 5. Consider the sample data in the table below, with expected frequencies shown in parentheses. Note that the first cell has an expected frequency of 3, which is less than 5, so the chi-square distribution should not be used.

Helmets and Facial Injuries in Bicycle Accidents

	Helmet Worn	No Helmet
Facial injuries received	2 (3)	13 (12)
All injuries nonfacial	6 (5)	19 (20)

Minitab procedure for Fisher's Exact Test

1. Enter the frequencies in column C1 with the corresponding row numbers in column C2 and the corresponding column numbers in column C3. Here are the Minitab entries for the above table. (The column labels of Frequency, Row, and Column were manually entered.) The first frequency of 2 is located in row 1 and column 1, as indicated by the entries in Minitab columns C2 and C3. The second frequency of 13 is in row 1 and column 2, and so on.

↓	C1	C2	C3
	Frequency	Row	Column
1	2	1	1
2	13	1	2
3	6	2	1
4	19	2	2

2. Select **Stat**, **Tables**, then **Cross Tabulation and Chi-Square**.

3. Enter C2 in the "For rows" box, enter C3 in the "For columns" box, and enter C1 in the "Frequencies are in" box.

4. Click on the **Other Stats** button and select "Fisher's exact test for 2×2 tables."

5. Click **OK** twice to get the resulting *P*-value. Here is the display that results from the above table:

```
Fisher's exact test: P-Value = 0.685661
```

Because the *P*-value is large, we fail to reject the null hypothesis that wearing a helmet and receiving facial injuries are independent. There isn't enough evidence to suggest that facial injuries are dependent on whether a helmet was worn.

CHAPTER 11: Goodness-of-Fit and Contingency Tables

11-1. ***Loaded Die*** The author drilled a hole in a die and filled it with a lead weight, then proceeded to roll it 200 times. Here are the observed frequencies for the outcomes of 1, 2, 3, 4, 5, and 6 respectively: 27, 31, 42, 40, 28, 32. Use a 0.05 significance level to test the claim that the outcomes are not equally likely.

Test statistic:__________ *P*-value:__________

Conclusion:____

____

Does it appear that the loaded die behaves differently than a fair die?

____

11–2. ***Flat Tire and Missed Class*** A classic tale involves four car-pooling students who missed a test and gave as an excuse a flat tire. On the makeup test, the instructor asked the students to identify the particular tire that went flat. If they really didn't have a flat tire, would they be able to identify the same tire? The author asked 41 other students to identify the tire they would select. The results are listed in the following table (except for one student who selected the spare). Use a 0.05 significance level to test the author's claim that the results fit a uniform distribution.

Tire	Left front	Right front	Left rear	Right rear
Number selected	11	15	8	6

Test statistic:__________ *P*-value:__________

Conclusion:____

____

What does the result suggest about the ability of the four students to select the same tire when they really didn't have a flat?

____

11-3. ***Births*** Records of randomly selected births were obtained and categorized according to the day of the week that they occurred (based on data from the National Center for Health Statistics). Because babies are unfamiliar with our schedule of weekdays, a reasonable claim is that births occur on the different days with equal frequency. Use a 0.01 significance level to test that claim.

Day	Sun	Mon	Tues	Wed	Thurs	Fri	Sat
Births	77	110	124	122	120	123	97

Test statistic:__________ *P*-value:__________

Conclusion:____

____

11-4. ***NYC Homicides*** For a recent year, the following are the numbers of homicides that occurred each month in New York City: 38, 30, 46, 40, 46, 49, 47, 50, 50, 42, 37, 37. Use a 0.05 significance level to test the claim that homicides in New York City are equally likely for each of the 12 months.

Test statistic:__________ *P*-value:__________

Conclusion:____

____

11–5. ***Measuring Pulse Rates*** According to one procedure used for analyzing data, when certain quantities are measured, the last digits tend to be uniformly distributed, but if they are estimated or reported, the last digits tend to have disproportionately more 0s or 5s. Refer to Data Set 1 in Appendix B of the textbook and use the last digits of the pulse rates of the 80 men and women. Those pulse rates were obtained as part of the National Health and Examination Survey. Test the claim that the last digits of 0, 1, 2, 3, . . . , 9 occur with the same frequency.

Test statistic:__________ *P*-value:__________

Conclusion:____

____

What can be inferred about the procedure used to obtain the pulse rates?

____

11-6. ***Testing a Normal Distribution*** In this experiment we will use Minitab's ability to generate normally distributed random numbers. We will then test the sample data to determine if they actually do fit a normal distribution.

a. Generate 1000 random numbers from a normal distribution with a mean of 100 and a standard deviation of 15. (IQ scores have these parameters.) Select **Calc**, then **Random Data**, then **Normal**.

b. Use **Data/Sort** to arrange the data in order.

c. Examine the sorted list and determine the frequency for each of the categories listed below. Enter those frequencies in the spaces provided. (The expected frequencies are found by assuming that the data are normally distributed.)

	Observed Frequency	Expected Frequency
Below 55:	____________	1
55-70:	____________	22
70-85:	____________	136
85-100:	____________	341
100-115:	____________	341
115-130:	____________	136
130-145:	____________	22
Above 145:	____________	1

d. Use Minitab to test the claim that the randomly generated numbers actually do fit a normal distribution with mean 100 and standard deviation 15.

Test statistic:__________ *P*-value:__________

Conclusion:____

____

11-7. ***Instant Replay in Tennis*** The table below summarizes challenges made by tennis players in the first U. S. Open that used the Hawk-Eye electronic instant replay system. Use a 0.05 significance level to test the claim that success in challenges is independent of the gender of the player. Does either gender appear to be more successful?

	Was the challenge to the call successful? Yes	No
Men	201	288
Women	126	224

Test statistic:__________ *P*-value:__________

Conclusion:____

____

11–8. ***Accuracy of Polygraph Tests*** The data in the accompanying table summarize results from tests of the accuracy of polygraphs (based on data from the Office of Technology Assessment). Use a 0.05 significance level to test the claim that whether the subject lies is independent of the polygraph indication.

	Polygraph Indicated Truth	Polygraph Indicated Lie
Subject actually told the truth	65	15
Subject actually told a lie	3	17

Test statistic:__________ *P*-value:__________

Conclusion:____

____

What do the results suggest about the effectiveness of polygraphs?

____

____

11–9. ***Global Warming Survey*** A Pew Research poll was conducted to investigate opinions about global warming. The respondents who answered yes when asked if there is solid evidence that the earth is getting warmer were then asked to select a cause of global warming. The results are given the table below. Use a 0.05 significance level to test the claim that the sex of the respondent is independent of the choice for the cause of global warming. Do men and women appear to agree, or is there a substantial difference?

	Human activity	Natural patterns	Don't know or refused to answer
Male	314	146	44
Female	308	162	46

Test statistic:__________ *P*-value:__________

Conclusion:____

____

11–10. ***Occupational Hazards*** Use the data in the table to test the claim that occupation is independent of whether the cause of death was homicide. The table is based on data from the U.S. Department of Labor, Bureau of Labor Statistics.

	Police	Cashiers	Taxi Drivers	Guards
Homicide	82	107	70	59
Cause of Death Other than Homicide	92	9	29	42

Test statistic:__________ *P*-value:__________

Conclusion:____

____

Does any particular occupation appear to be most prone to homicides? If so, which one?

____

How are the results affected if the order of the rows is switched?

____

How are the results affected by the presence of an outlier? If we change the first entry from 82 to 8200, are the results dramatically affected?

____

11-11. ***Fisher's Exact Test*** Refer to Experiment 11-8 in this manual/workbook. Repeat that experiment by using Fisher's exact test instead of using the approximating chi-square distribution. Enter the results below.

P-value obtained by using the approximating chi-square distribution: __________

P-value obtained by using Fisher's exact test:

____

Does the use of the Fisher's exact test have much of an effect on the *P*-value?

____

____

11-12. ***Fisher's Exact Test*** The U. S. Supreme Court considered a case involving the exam for firefighter lieutenant in the city of New Haven, CT. Results from the exam are shown in the table below. Is there sufficient evidence to support the claim that results from the test should be thrown out because they are discriminatory? Use a 0.01 significance level.

	Passed	Failed
White Candidates	17	16
Minority Candidates	9	25

P-value obtained by using the approximating chi-square distribution: __________

Conclusion (based on results from using the approximating chi-square distribution):

____

P-value obtained by using Fisher's exact test: __________

Conclusion (based on results from using Fisher's exact test):

____

Does the use of the Fisher's exact test have much of an effect on the *P*-value?

____

____

12

Analysis of Variance

12-1 One-Way Analysis of Variance

One–way analysis of variance is used to test the claim that three or more populations have the same mean. When the Triola textbook discusses one-way analysis of variance, it is noted that the term "one-way" is used because the sample data are separated into groups according to one characteristic or "factor". In Table 12-1 the factor is the blood lead level (low, medium, high).

Table 12-1 Performance IQ Scores of Children

Low Blood Lead Level

85 90 107 85 100 97 101 64 111 100 76 136 100 90 135 104 149 99 107 99
113 104 101 111 118 99 122 87 118 113 128 121 111 104 51 100 113 82 146 107
83 108 93 114 113 94 106 92 79 129 114 99 110 90 85 94 127 101 99 113
80 115 85 112 112 92 97 97 91 105 84 95 108 118 118 86 89 100

Medium Blood Lead Level

78 97 107 80 90 83 101 121 108 100 110
111 97 51 94 80 101 92 100 77 108 85

High Blood Lead Level

93 100 97 79 97 71 111 99 85 99 97 111 104 93 90 107 108 78 95 78 86

Because the calculations are very complicated, the Triola textbook emphasizes the interpretation of results obtained by using software, so Minitab is very suitable for this topic. We should understand that a small *P*-value (such as 0.05 or less) leads to rejection of the null hypothesis of equal means. ["If the *P* (value) is low, the null must go."] With a large *P*-value (such as greater than 0.05), fail to reject the null hypothesis of equal means.

For the data in the above table, the claim of equal means leads to these hypotheses:

H_0: $\mu_1 = \mu_2 = \mu_3$

H_1: At least one of the population means is different from the others.

With Minitab Release 16 or later, you could use the new "Assistant" feature by clicking on the main menu item of **Assistant.** Select **Hypothesis Tests,** then select **One-Way ANOVA.** Use of the Assistant feature yields four screens of results that include much helpful information. The following procedure describes another way that Minitab can be used to test the claim that the different samples come from populations with the same mean.

Minitab Procedure for One-Way Analysis of Variance

1. Enter the original sample values in columns (or open a Minitab worksheet that contains the columns of data).
2. Select **Stat** from the main menu.
3. Select the subdirectory item of **ANOVA**.
4. Select the option of **Oneway (Unstacked)**. The term "unstacked" means that the data are not all stacked in one single column; they are listed in their individual and separate columns.
5. In the dialog box, enter the column names (C1 C2 C3) in the box labeled as "Responses (in separate columns)." Click **OK**.

If you use the above steps with the performance IQ scores from the above table, the Minitab results will appear as shown below.

One-way ANOVA: Low, Medium, High

```
Source   DF     SS    MS      F      P
Factor    2   2023  1011   4.07  0.020
Error   118  29314   248
Total   120  31337

S = 15.76   R-Sq = 6.45%   R-Sq(adj) = 4.87%

                              Individual 95% CIs For Mean Based on
                              Pooled StDev
Level    N    Mean  StDev  -----+---------+---------+---------+----
Low     78  102.71  16.79                         (------*------)
Medium  22   94.14  15.48  (------------*-------------)
High    21   94.19  11.37  (------------*-------------)
                           -----+---------+---------+---------+----
                               90.0      95.0     100.0     105.0

Pooled StDev = 15.76
```

The P-value of 0.020 indicates that there is sufficient sample evidence to warrant rejection of the null hypothesis that $\mu_1 = \mu_2 = \mu_3$. The test statistic of $F = 4.07$ is also provided, as are the values of the SS and MS components. The lower portion of the Minitab display also includes the individual sample means, standard deviations, and graphs of the 95% confidence interval estimates of each of the three population means. These confidence intervals are computed by the same methods used in Chapter 7, except that a pooled standard deviation is used instead of the individual sample standard deviations. (The last entry in the Minitab display shows that the pooled standard deviation is 15.76.)

Note that the Minitab display uses the term *factor* instead of *treatment*, so the value of SS(treatment) is 2023 and MS(treatment) = 1011.

12-2 Two-Way Analysis of Variance

Two-way analysis of variance involves *two* factors, such as type of gender (male or female) and blood lead level (low, medium high) in Table 12-3. The two–way analysis of variance procedure requires that we test for (1) an interaction effect between the two factors; (2) an effect from the row factor; (3) an effect from the column factor.

Table 12-3	**Measures of Performance IQ**		
	Blood Lead Level		
	Low	**Medium**	**High**
Male	85	78	93
	90	107	97
	107	90	79
	85	83	97
	100	101	111
Female	64	97	100
	111	80	71
	76	108	99
	136	110	85
	99	97	93

Entering the Sample Data The first obstacle to overcome is to understand a somewhat awkward method of entering sample data. When entering the values in the above table, we must somehow keep track of the location of each data value, so here is the procedure to be used:

1. Enter *all* of the sample data in column C1.
2. Enter the corresponding row numbers in column C2.
3. Enter the corresponding column numbers in column C3.

This particular format can be confusing at first, so stop and try to recognize the pattern. Because the first value of 85 from the above table is in the first row and first column, its row identification is 1 and its column identification is 1. The entry of 90 is also in the first row and first column, so its row identification number is 1 and its column identification number is 1. (The sample values of 85, 90, 107, 85, 100 are all considered to be in the first row (male) and the first column (low). The value of 78 is in the first row and second column.

Also, it's wise to name column C1 as "IQ," name column C2 as "SEX", and name column C3 "LEVEL" so that the columns have meaningful names. Here is an illustration of the pattern used to enter the sample data in the above table. The Minitab display includes only the first 12 values from Table 12-3.

Table 12-3	**Measures of Performance IQ**		
	Blood Lead Level		
	Low	**Medium**	**High**
Male	85	78	93
	90	107	97
	107	90	79
	85	83	97
	100	101	111
Female	64	97	100
	111	80	71
	76	108	99
	136	110	85
	99	97	93

↓	C1	C2	C3
	IQ	SEX	LEVEL
1	85	1	1
2	90	1	1
3	107	1	1
4	85	1	1
5	100	1	1
6	78	1	2
7	107	1	2
8	90	1	2
9	83	1	2
10	101	1	2
11	93	1	3
12	97	1	3

The procedure for obtaining a Minitab display for two-way analysis of variance is as follows.

Minitab Procedure for Two-Way Analysis of Variance

1. Enter all of the sample values in column C1 (as shown above).

2. Enter the corresponding row numbers in column C2 (as shown above).

3. Enter the corresponding column numbers in column C3 (as shown above).

4. Select **Stat**, then **ANOVA**, then **Twoway**.

5. You should now see a dialog box like the one shown below.

   Make these entries in the dialog box:

   - Enter C1 (or its column label) in the box labeled Response.
   - Enter C2 (or its column label) in the box labeled Row Factor.
   - Enter C3 (or its column label) in the box labeled Column Factor.
   - Click **OK**.

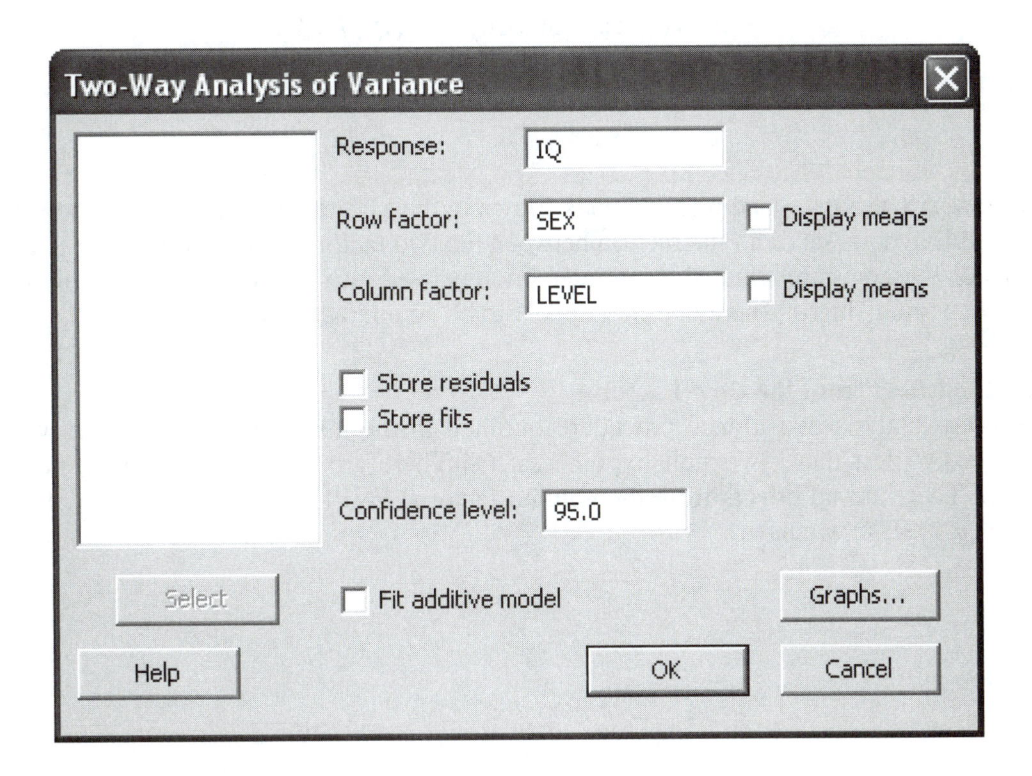

Using the data in Table 12-3 will result in the Minitab display shown below.

Two-way ANOVA: IQ versus SEX, LEVEL

Source	DF	SS	MS	F	P
SEX	1	17.63	17.633	**0.07**	**0.791**
LEVEL	2	48.80	24.400	**0.10**	**0.906**
Interaction	2	211.47	105.733	**0.43**	**0.655**
Error	24	5886.40	245.267		
Total	29	6164.30			

S = 15.66   R-Sq = 4.51%   R-Sq(adj) = 0.00%

The Triola textbooks (excluding *Essentials of Statistics*) describe the interpretation of the preceding Minitab display.

1. Test for Interaction

We begin by testing the null hypothesis that there is no *interaction* between the two factors of SEX and LEVEL. Using the above Minitab results, we calculate the following test statistic.

$$F = \frac{\text{MS}(\text{interaction})}{\text{MS}(\text{error})} = \frac{105.733}{245.267} = 0.43$$

Interpretation: The corresponding *P*-value is shown in the Minitab display as 0.655, so we fail to reject the null hypothesis of no interaction between the two factors. It does not appear that the performance IQ scores are affected by an interaction between sex (male, female) and blood lead level (low, medium, high). There does not appear to be an interaction effect.

2. Test for Effect from the Row Factor

Our two-way analysis of variance procedure outlined in the textbook indicates that we should now proceed to test these two null hypotheses: (1) There are no effects from the row factor (SEX); (2) There are no effects from the column factor (LEVEL). For the test of an effect from the row factor (SEX), we have

$$F = \frac{\text{MS}(\text{sex})}{\text{MS}(\text{error})} = \frac{17.633}{245.267} = 0.07$$

Conclusion: The corresponding *P*-value is shown in the Minitab display as 0.791. Because that *P*-value is greater than the significance level of 0.05, we fail to reject the null hypothesis of no effects from sex. That is, performance IQ scores do not appear to be affected by the sex of the subject.

3. Test for Effect from the Column Factor
For the test of an effect from SIZE, we have

$$F = \frac{\text{MS}(\text{lead level})}{\text{MS}(\text{error})} = \frac{24.400}{245.267} = 0.10$$

Conclusion: The corresponding *P*-value is shown in the Minitab display as 0.906. Because that *P*-value is greater than the significance level of 0.05, we fail to reject the null hypothesis of no effects from lead level. Performance IQ scores do not appear to be affected by whether the lead exposure is low, medium, or high.

Special Case: One Observation Per Cell
The Triola textbook (excluding *Essentials of Statistics*) includes a subsection describing the special case in which there is only one sample value in each cell. Here's how we proceed when there is one observation per cell: *If it seems reasonable to assume (based on knowledge about the circumstances) that there is no interaction between the two factors, make that assumption and then proceed as before to test the following two hypotheses separately:*

H_0: There are no effects from the row factor.

H_0: There are no effects from the column factor.

To use Minitab, simply apply the same procedure described earlier. The Minitab results will not include values for an interaction, but the other necessary values are provided.

CHAPTER 12 EXPERIMENTS: Analysis of Variance

12-1. **Clancy, Rowling, Tolstoy Readability** Pages were randomly selected by the author from *The Bear and the Dragon* by Tom Clancy, *Harry Potter and the Sorcerer's Stone* by J. K. Rowling, and *War and Peace* by Leo Tolstoy. The Flesch Reading Ease scores for those pages are listed below. Use a 0.05 significance level to test the claim that the three samples are from populations with the same mean. Do the books appear to have different reading levels of difficulty?

Clancy	58.2	73.4	73.1	64.4	72.7	89.2	43.9	76.3	76.4	78.9	69.4	72.9
Rowling	85.3	84.3	79.5	82.5	80.2	84.6	79.2	70.9	78.6	86.2	74.0	83.7
Tolstoy	69.4	64.2	71.4	71.6	68.5	51.9	72.2	74.4	52.8	58.4	65.4	73.6

SS(treatment): ________ MS(treatment): ________ Test statistic *F*: ________

SS(error): ________ MS(error): ________ *P*-value: ________

SS(total): ________

Conclusion: ____

____

12–2. **Poplar Tree Weights** Weights (kg) of poplar trees were obtained from trees planted in a sandy and dry region. The trees were given different treatments identified in the table below. The data are from a study conducted by researchers at Pennsylvania State University, and the data were provided by Minitab, Inc. Use a 0.05 significance level to test the claim that the four treatment categories yield poplar trees with the same mean weight. Is there a treatment that appears to be most effective in the sandy and dry region?

No Treatment	Fertilizer	Irrigation	Fertilizer and Irrigation
1.21	0.94	0.07	0.85
0.57	0.87	0.66	1.78
0.56	0.46	0.10	1.47
0.13	0.58	0.82	2.25
1.30	1.03	0.94	1.64

SS(treatment): ________ MS(treatment): ________ Test statistic *F*: ________

SS(error): ________ MS(error): ________ *P*-value: ________

SS(total): ________

Conclusion: ____

12–3. ***Archeology: Skull Breadths from Different Epochs*** The values in the table are measured maximum breadths of male Egyptian skulls from different epochs (based on data from Ancient Races of the Thebaid, by Thomson and Randall-Maciver). Changes in head shape over time suggest that interbreeding occurred with immigrant populations. Use a 0.05 significance level to test the claim that the different epochs do not all have the same mean.

4000 B.C.	1850 B.C.	150 A.D.
131	129	128
138	134	138
125	136	136
129	137	139
132	137	141
135	129	142
132	136	137
134	138	145
138	134	137

SS(treatment): ________ MS(treatment): ________ Test statistic F: ________

SS(error): ________ MS(error): ________ P-value: ________

SS(total): ________

Conclusion:____

____

12–4. ***Mean Weights of M&Ms*** Refer to the M&M Data Set 20 in Appendix B from the textbook. (The Minitab worksheet name is M&M.) At the 0.05 significance level, test the claim that the mean weight of M&Ms is the same for each of the six different color populations.

SS(treatment): ________ MS(treatment): ________ Test statistic F: ________

SS(error): ________ MS(error): ________ P-value: ________

SS(total): ________

Conclusion:____

If it is the intent of Mars, Inc., to make the candies so that the different color populations have the same mean weight, do these results suggest that the company has a problem requiring corrective action?

____

12–5. ***Nicotine in Cigarettes*** Refer to Data Set 10 in Appendix B and use the amounts of nicotine (mg per cigarette) in the king size cigarettes, the 100 mm menthol cigarettes, and the 100 mm non-menthol cigarettes. (The Minitab worksheet name is CIGARET.) The king size cigarettes are non-filtered, non-menthol, and non-light. The 100 mm menthol cigarettes are filtered and non-light. The 100 mm non-menthol cigarettes are filtered and non-light. Use a 0.05 significance level to test the claim that the three categories of cigarettes yield the same mean amount of nicotine. Given that only the king size cigarettes are not filtered, do the filters appear to make a difference?

SS(treatment): ________ MS(treatment): ________ Test statistic F: ________

SS(error): ________ MS(error): ________ P-value: ________

SS(total): ________

Conclusion:____

12–6. ***Tar in Cigarettes*** Refer to Data Set 10 in Appendix B and use the amounts of tar (mg per cigarette) in the three categories of cigarettes described in Experiment 12-5. (The Minitab worksheet name is CIGARET.) Use a 0.05 significance level to test the claim that the three categories of cigarettes yield the same mean amount of tar. Given that only the king size cigarettes are not filtered, do the filters appear to make a difference?

SS(treatment): ________ MS(treatment): ________ Test statistic F: ________

SS(error): ________ MS(error): ________ P-value: ________

SS(total): ________

Conclusion:____

12-7. ***Simulations*** Use Minitab to randomly generate three different samples of 500 values each. (Select **Calc, Random Data, Normal**.) For the first two samples, use a normal distribution with a mean of 100 and a standard deviation of 15. For the third sample, use a normal distribution with a mean of 105 and a standard deviation of 15. We know that the three populations have different means, but do the methods of analysis of variance allow you to conclude that the means are different? Explain.

____

____

____

12–8. ***Pulse Rate*** The following table lists pulse rates obtained from Data Set 1 in Appendix B of the textbook. Use a 0.05 significance level and apply the methods of two-way analysis of variance. What do you conclude?

	Under 30 Years of Age	Over 30 Years of Age
Female	78 104 78 64 60 98 82 98 90 96	76 76 72 66 72 78 62 72 74 56
Male	60 80 56 68 68 74 74 68 62 56	46 70 62 66 90 80 60 58 64 60

Are pulse rates affected by an interaction between gender and age? Explain.

____

Are pulse rates affected by gender? Explain.

____

Are pulse rates affected by age? Explain.

____

____

12-9. ***Smoking, Body Temperature, Gender*** The table below lists body temperatures obtained from randomly selected subjects (based on Data Set 3 in Appendix B). The temperatures are categorized according to gender and whether the subject smokes. Using a 0.05 significance level, test for an interaction between gender and smoking, test for an effect from gender, and test for an effect from smoking. What do you conclude?

	Smokes	Does not smoke
Male	8.4 98.4 99.4 98.6	98.0 98.0 98.8 97.0
Female	8.8 98.0 98.7 98.4	97.7 98.0 98.2 99.1

Are body temperatures affected by an interaction between sex and smoking? Explain.

____

Are body temperatures affected by sex? Explain.

____

Are body temperatures affected by smoking? Explain.

____

____

13

Nonparametric Statistics

13-1 Ranking Data

This chapter describes how Minitab can be used for the nonparametric methods presented in the Triola textbook. The sections of this chapter correspond to those in the textbook. The textbook introduces some basic principles of nonparametric methods, but it also describes a procedure for converting data into their corresponding *ranks*. Here is the Minitab procedure for using data in column C1 to create a column C2 that consists of the corresponding ranks.

Converting Data to Ranks

1. Enter the data values in column C1.

2. Select **Data** from the main menu.

3. Select **Rank** from the subdirectory.

4. In the dialog box, make these entries:

- For the box labeled "Rank data in," enter C1.
- For the box labeled "Store ranks in," enter C2.
- Click **OK**.

As an example, consider the values of 5, 3, 40, 10, 12, and 12. They are converted to the corresponding ranks of 2, 1, 6, 3, 4.5, and 4.5. Enter the values of 5, 3, 40, 10, 12, and 12 in column C1, follow the above steps, and column C2 will contain the ranks of 2, 1, 6, 3, 4.5, and 4.5 as shown below.

↓	C1	C2
1	5	2.0
2	3	1.0
3	40	6.0
4	10	3.0
5	12	4.5
6	12	4.5

Ties: Note that Minitab handles ties as described in the Triola textbook. See the above display that includes ties in the ranks that correspond to the values of 12 and 12. Ties occur with the corresponding ranks of 4 and 5, so Minitab assigns a rank of 4.5 to each of those values.

13-2 Sign Test

The Triola textbook (excluding *Essentials of Statistics*) includes the following definition.

> ***Definition***
> The **sign test** is a nonparametric (distribution-free) test that uses plus and minus signs to test different claims, including:
> 1. Claims involving matched pairs of sample data
> 2. Claims involving nominal data with two categories
> 3. Claims about the median of a single population

Minitab makes it possible to work with all three of the above cases. Let's consider the matched data in the table below, from Example 2 in Section 13-2 of the textbook. (The data are matched, because each pair of values is from the same flight.)

Taxi-Out Times and Taxi-In Times for American Airlines Flight 21

Taxi-out time	13	20	12	17	35	19	22	43	49	45	13	23
Taxi-in time	13	4	6	21	29	5	27	9	12	7	36	12
Sign of difference	0	+	+	−	+	+	−	+	+	+	−	+

Given such paired data, we can use Minitab to apply the sign test to test the claim of no difference.

Minitab Procedure for Sign Test with Matched Data

1. Enter the values of the first variable in column C1.

2. Enter the values of the second variable in column C2.

3. Create column C3 equal to the values of C1 – C2. Accomplish this using either of the following two approaches.

   Session window: Click on the session window at the top portion of the screen, then click on **Editor**, then click on **Enable Commands**. Enter the command:
   LET C3 = C1 – C2.

   Calculator: Click on **Calc**, then **Calculator**. In Minitab's calculator, enter C3 for the "Store result in variable" box, and enter the expression C1 – C2 in the "expression" box. Click **OK** when done.

4. Select **Stat** from the main menu.

5. Select **Nonparametrics** from the subdirectory.

6. Select the option of **1-Sample Sign**.

7. Make these entries in the resulting dialog box:

- For the box labeled Variables, enter C3 (which contains the differences obtained when column C2 is subtracted from column C1).

- Click on the small circle for **Test median** and, in the adjacent box, enter 0 for a test of the claim that the median is equal to zero.

- In the box labeled Alternative, select **not equal,** so that the alternative hypothesis is that the statement that the median is not equal to zero.

- Click **Ok**.

Here are the Minitab results for the data in the above table:

```
Sign test of median =  0.00000 versus not = 0.00000

      N  Below  Equal  Above         P  Median
C3   12      3      1      8    0.2266   8.500
```

The above results show that there are 12 pairs of data, there are 3 differences that are below zero (negative sign), one of the differences is equal to zero, and 8 of the differences are above zero (positive sign). The *P*-value is 0.2266. In the textbook, we would use a decision criterion that involves a comparison of the test statistic $x = 3$ (the less frequent sign) and a critical value found in Table A-7. Because Minitab displays the *P*-value, we can use the *P*-value approach to hypothesis testing. Because the *P*-value of 0.2266 is greater than the significance level of $\alpha =$ 0.05, there is not sufficient evidence to warrant rejection of the claim that the median of the differences is equal to 0. There does not appear to be significant difference between the taxi-out times and the taxi-in times.

Sign test with nominal data: In some cases, the data for the sign test do not consist of values of the type given in the table above. For example, suppose that a test of a gender-selection method results in 945 babies, including 879 girls and 66 boys (as in Example 2 in the textbook). Because the data are at the nominal level of measurement, we could enter them as –1's (for girls) and 1's (for boys). *Shortcut*: Instead of entering –1 a total of 879 times and 1 a total of 66 times, we can enter the following commands in the session window to achieve the same result. (Click on the session window at the top portion of the screen, then click on **Editor**, then click on **Enable Commands**. Enter these commands to create a column consisting of –1 repeated 879 times and 1 repeated 66 times.)

```
SET C3
879(-1), 66(1)
END
```

Now follow the procedure given earlier (beginning with step 3) for conducting the sign test, and the results will be as shown below.

```
Sign test of median =  0.00000 versus not = 0.00000

       N  Below  Equal  Above        P  Median
C3   945    879      0     66  0.0000  -1.000
```

The resulting *P*-value of 0.0000 suggests that we reject the null hypothesis of equality of girls and boys. There is sufficient evidence to warrant rejection of the claim that girls and boys are equally likely.

Sign test for testing a claimed value of the median: The textbook also includes an example involving body temperature data. (See Data Set 3 in Appendix B of the textbook for the body temperatures for 12:00 AM on day 2.) We can apply the sign test to test the claim that the *median* is less than 98.6°. Use the CD included with this book to open the Minitab worksheet **BODYTEMP** (or manually enter the 106 body temperatures) and follow the procedure for doing the sign test, but make these changes in the dialog box entries of Step 6: Enter 98.6 in the box adjacent to "Test median," and select "less than" in the box labeled Alternative. The Minitab results will be as shown below.

```
Sign test of median = 98.60 versus  <  98.60

                  N  Below  Equal  Above          P      Median
C1              106     68     15     23     0.0000       98.40
```

The above Minitab display shows that among the 106 body temperatures, 68 are below 98.6, 15 are equal to 98.6, and 23 are above 98.6. The *P*-value of 0.0000 causes us to reject the null hypothesis that the median is at least 98.6. There appears to be sufficient evidence to support the claim that the median body temperature is less than 98.6° F.

13-3 Wilcoxon Signed-Ranks Test

The Triola textbook (excluding *Essentials of Statistics*) describes the Wilcoxon signed-ranks test, and the following definition is given.

Definition

The **Wilcoxon signed-ranks test** is a nonparametric test that uses ranks for these applications:

1. Testing a claim that the population of matched pairs has the property that the matched pairs have differences with median equal to zero

2. Testing a claim that a single population of individual values has median equal to some claimed value

The textbook makes this important point: The Wilcoxon signed-ranks test and the sign test can both be used with sample data consisting of matched pairs, but the sign test uses only the *signs* of the differences and not their actual magnitudes (how large the numbers are). The Wilcoxon signed-ranks test uses *ranks*, so the magnitudes of the differences are taken into account. Because the Wilcoxon signed-ranks test incorporates and uses more information than the sign test, it tends to yield conclusions that better reflect the true nature of the data. First we describe the Minitab procedure for conducting a Wilcoxon signed-ranks test, then we illustrate it with an example.

Minitab Procedure for the Wilcoxon Signed-Ranks Test

1. Enter the values of the first variable in column C1.

2. Enter the values of the second variable in column C2.

3. Create column C3 equal to the values of C1 – C2. Accomplish this using either of the following two approaches.

   Session window: Click on the session window at the top portion of the screen, then click on **Editor**, then click on **Enable Commands**. Enter the command:
   LET C3 = C1 – C2.

   Calculator: Click on **Calc**, then **Calculator**. In Minitab's calculator, enter C3 for the "Store result in variable" box, and enter the expression C1 – C2 in the "expression" box. Click **OK** when done.

4. Select **Stat** from the main menu.

5. Select **Nonparametrics** from the subdirectory.

(*continued*)

6. Select the option of **1-Sample Wilcoxon**.

7. Make these entries in the resulting dialog box:

- **Variables:** For the box labeled Variables, enter C3 (which contains the differences obtained when column C2 is subtracted from column C1).

- **Test Median:** Click on the button for "Test median," and in the adjacent box, enter 0 for a test of the claim that the median is equal to zero.

- **Alternative:** In the box labeled Alternative, select "not equal" for a test of the claim that the median is equal to zero (so that the alternative hypothesis is that the median is not equal to zero).

- Click **OK**.

Using the same matched data from the table in Section 13-2 of this manual/worksheet (or Example 1 in Section 13-3 of the textbook), the Minitab results for the Wilcoxon signed-ranks test will be as shown below.

Wilcoxon Signed Rank Test: C3

```
Test of median = 0.000000 versus median not = 0.000000

             N for   Wilcoxon            Estimated
        N     Test  Statistic        P      Median
C3    945      945    31218.0    0.000      -1.000
```

From these results we see that the *P*-value is 0.000. Because the *P*-value of 0.000 is less than the significance level of 0.05, we reject the null hypothesis that the matched pairs have differences with a median equal to 0. There does appear to be a significant difference between the taxi-in times and the corresponding taxi-out times. (Note that the sign test in Section 13-2 led to the conclusion of no difference. By using only positive and negative signs, the sign test did not use the magnitudes of the differences, but the Wilcoxon signed-ranks test was more sensitive to those magnitudes through its use of ranks.)

13-4 Wilcoxon Rank-Sum Test

The Triola textbook (excluding *Essentials of Statistics*) discusses the Wilcoxon rank-sum test and includes the following definition.

Definition

The **Wilcoxon rank-sum test** is a nonparametric test that uses ranks of sample data from two independent populations to test this null hypothesis: H_0: The two independent samples come from populations with equal medians. (The alternative hypothesis H_1 can be any one of the following three possibilities: The two populations have different medians, or the first population has a median *greater than* the median of the second population, or the first population has a median *less than* the median of the second population.)

The Wilcoxon rank-sum test described in the textbook is equivalent to the Mann-Whitney *U* test, so use Minitab's Mann-Whitney procedure for testing the claim that two independent samples come from populations with populations with equal medians.

Minitab Procedure for the Mann-Whitney Test
(Equivalent to the Wilcoxon Rank-Sum Test)

1. Enter the two sets of sample data in columns C1 and C2.
2. Select **Stat** from the main menu.
3. Select **Nonparametrics**.
4. Select **Mann-Whitney**.
5. Make these entries in the dialog box:
    - Enter C1 for the first sample and enter C2 for the second sample.
    - Enter the confidence level in the indicated box. (A confidence level of 95.0 corresponds to a significance level of $\alpha = 0.05$.)
    - Select the desired form of the alternative hypothesis.

Shown below are pulse rates of males and females (from Data Set 1 in Appendix B). (Ranks are shown in parentheses.) We want to use a 0.05 significance level to test the claim that males and females have the same median pulse rate (as in Example 1 in the textbook). Also shown are the Minitab results that include a *P*-value of 0.2166. Based on the available sample data, it appears that there is not sufficient evidence to warrant rejection of the claim that males and females have pulse rates with the same median.

Table 13-5 Pulse Rates
(Ranks in parentheses)

Males	Females
60 (**4.5**)	78 (**14**)
74 (**11**)	80 (**17**)
86 (**19**)	68 (**9**)
54 (**1**)	56 (**2.5**)
90 (**20.5**)	76 (**12**)
80 (**17**)	78 (**14**)
66 (**7**)	78 (**14**)
68 (**9**)	90 (**20.5**)
68 (**9**)	96 (**22**)
56 (**2.5**)	60 (**4.5**)
80 (**17**)	98 (**23**)
62 (**6**)	
$n_1 = 12$	$n_2 = 11$
$R_1 = 123.5$	$R_2 = 152.5$

Mann-Whitney Test and CI: Males, Females

```
           N  Median
Males     12   68.00
Females   11   78.00

Point estimate for ETA1-ETA2 is -10.00
95.5 Percent CI for ETA1-ETA2 is (-20.00,4.01)
W = 123.5
Test of ETA1 = ETA2 vs ETA1 not = ETA2 is significant at 0.2184
The test is significant at 0.2166 (adjusted for ties)
```

13-5 Kruskal-Wallis Test

The Triola textbook (excluding *Essentials of Statistics*) discusses the Kruskal-Wallis test and includes this definition.

Definition

The **Kruskal-Wallis Test** (also called the ***H* test**) is a nonparametric test that uses ranks of simple random samples from three or more independent populations to test the null hypothesis that the populations have the same median. (The alternative hypothesis is the claim that the populations have medians that are not all equal.)

Caution: Using Minitab for a Kruskal-Wallis test is somewhat tricky, because we must enter all of the data in *one column*, then enter the corresponding identifiers in another column. For an example, see the sample data in the table below left (from Example 1 in Section 13-5 of the textbook), and see the format required by Minitab at the right. The columns at the right show only the first 10 data values, but all of the data must be listed with the same pattern.

Performance IQ Scores

Low Lead Level	Medium Lead Level	High Lead Level
85	78	93
90	97	100
107	107	97
85	80	79
100	90	97
97	83	
101		
64		

Stacked Data

85	Low
90	Low
107	Low
85	Low
100	Low
97	Low
101	Low
64	Low
78	Medium
97	Medium

Stacking Data If the data are listed in separate columns and you want to stack them as shown above, use **Data/Stack** to combine them into one single column. After selecting **Data/Stack**, you will get a "Stack Columns" dialog box. Enter the columns to be stacked, identify the column where the stacked data will be listed, and identify the column where the corresponding "subscripts" (identifiers) will be listed.

After the sample data have been arranged in a single column with identifiers in another column, we can proceed to use Minitab as follows.

Procedure for the Kruskal-Wallis Test

1. Enter the data values for all categories in one column, and enter the corresponding identification numbers or names in another column.

2. Select **Stat** from the main menu.

3. Select **Nonparametrics**.

4. Select **Kruskal-Wallis**.

5. Make these entries in the dialog box:

   - **Response:** For Response, enter the column with all of the sample data combined.

   - **Factor:** For Factor, enter the column of identification numbers or names.

   - Click **OK**.

Here are the Minitab results obtained by using the Kruskal-Wallis test with the data in the preceding table:

Kruskal-Wallis Test: IQ versus Lead

```
Kruskal-Wallis Test on IQ

Lead      N  Median  Ave Rank      Z
High      5   97.00      10.7   0.32
Low       8   93.50      10.8   0.50
Medium    6   86.50       8.4  -0.83
Overall  19              10.0

H = 0.69  DF = 2  P = 0.707
H = 0.70  DF = 2  P = 0.704  (adjusted for ties)
```

Important elements of the Minitab display include the test statistic of $H = 0.70$ and the P–value of 0.704. We fail to reject the null hypothesis (the samples come from populations with the same median) only if the P–value is small (such as 0.05 or less). Assuming a 0.05 significance level in this case, we fail to reject the null hypothesis of equal medians. The population medians do not appear to be different.

13-6 Rank Correlation

The Triola textbook introduces *rank correlation*, which uses ranks in a procedure for determining whether there is some relationship between two variables.

> ***Definition***
>
> The **rank correlation test** (or **Spearman's rank correlation test**) is a nonparametric test that uses ranks of sample data consisting of matched pairs. It is used to test for an association between two variables.

First we describe the Minitab procedure, then we illustrate it with an example.

Minitab Procedure for Rank Correlation

1. Enter the paired data in columns C1 and C2.

2. If the data are already ranks, go directly to step 3. If the data are not already ranks, convert them to ranks by first using Minitab's **Rank** feature. (See Section 13-1 of this manual/workbook.) To convert the data to ranks, make these entries in the Rank dialog box:

    - Enter C1 in the box labeled Rank the data in.
    - Enter C1 in the box labeled Store ranks in.
    - Click **OK**.
    - Enter C2 in the box labeled Rank the data in.
    - Enter C2 in the box labeled Store ranks in.
    - Click **OK**.

3. Select **Stat** from the main menu.

4. Select **Basic Statistics**.

5. Select **Correlation**.

6. Enter C1 C2 in the dialog box, then click **OK**.

Caution: When working with data having ties among ranks, the rank correlation coefficient r_s can be calculated using Formula 10-1. Technology can be used instead of manual calculations with formula 10-1, but the displayed *P*-values for linear correlation do not apply to the methods of rank correlation. *Do not use P-values from linear correlation for methods of rank correlation.*

Consider the sample data in the table below (from Example 1 in Section 13-6 of the textbook). The measures of overall quality do not consist of ranks, so we convert them to ranks using Minitab's **Rank** feature, as described in Step 2 above.

Overall Quality Scores and Prices of LCD Televisions

Quality Rank	1	2	3	4	5	6	7
Price (dollars)	1900	1200	1300	2000	1700	1400	2700

After replacing the measures of overall quality with their corresponding ranks, we get the table shown below.

***Ranks* of Overall Quality Scores and Prices of LCD Televisions**

Quality Rank	1	2	3	4	5	6	7
Price *Rank*	5	1	2	6	4	3	7

We now have a table in which both rows are ranks, so we can proceed with Step 3 above. After continuing with Step 3 and the remaining steps, we obtain the following Minitab results.

Correlations: Quality, Price

```
Pearson correlation of Quality and Price = 0.429
```

The Minitab display shows that the rank correlation coefficient is $r_s = 0.429$. *Ignore a displayed P-value.*(Minitab will display a *P*-value of 0.337, but the correct *P*-value is 0.354.) Instead, refer to Table A-9 in the textbook to find the critical values of −0.786 and 0.786. Because $r_s = 0.429$ falls between the critical values, we fail to reject the null hypothesis. There is not sufficient evidence to support a claim of a correlation between quality and price.

13-7 Runs Test for Randomness

The Triola textbook (excluding *Essentials of Statistics*) discusses the runs test for randomness and includes these definitions.

> ***Definitions***
>
> After characterizing each data value as one of two separate categories, a **run** is a sequence of data having the same characteristic; the sequence is preceded and followed by data with a different characteristic or by no data at all.
>
> The **runs test** uses the number of runs in a sequence of sample data to test for randomness in the order of the data.

When listing sequences of data to be used for the runs test, the textbook uses both numerical data and qualitative data, but Minitab works with *numerical* data only, so enter such a sequence using numbers instead of letters. As an illustration, consider the winners of the NBA basketball championship game, with W denoting a winner from the Western Conference and E denoting a winner from the Eastern Conference (as in Example 1 in Section 13-7 of the textbook). We get the following sequence of E's and W's, which we then convert to 0's and 1's as shown below.

```
E  E  W  W  W  W  W  E  W  E  W  E  W  W  W
0  0  1  1  1  1  1  0  1  0  1  0  1  1  1
```

Minitab Procedure for the Runs Test for Randomness

1. In column C1, enter the sequence of *numerical* data. (If the sequence of data consists of two qualitative characteristics, represent the sequence using 0's and 1's.)

2. Select **Stat** from the main menu.

3. Select **Nonparametrics**.

4. Select **Runs** Test.

5. Make these entries in the dialog box:

   - Enter C1 in the box for variables.

   - Click on the button for **Above and below**, and enter a number in the adjacent box. (Choose a number appropriate for the test. If using 0's and 1's, enter 0.5.)

   - Click **OK**.

Example 3 in the textbook refers to a sequence of 107 genders (M and F) obtained from Data Set 3 in Appendix B of the textbook. We want to use a 0.05 significance level to test the claim that the sequence of genders is random. After denoting each M by the number 1 and denoted each F by the number 0, we get a column of 107 numbers. The runs test results are as shown below (based on testing for randomness above and below 0.5).

Runs Test: Gender

```
Runs test for Gender

Runs above and below K = 0.5

The observed number of runs = 25
The expected number of runs = 26.7944
92 observations above K, 15 below
P-value = 0.465
```

The above results show that the *P*-value is 0.465, there are 25 runs, there are 92 values above 0.5 (or 92 males), and there are 15 values below 0.5 (or 15 females). Because the *P*-value is greater than the significance level of 0.05, we fail to reject the null hypothesis of randomness.

CHAPTER 13 EXPERIMENTS: Nonparametric Statistics

In Experiments 13–1 through 13–7, use Minitab's **sign test** *program.*

13-1. ***Oscar Winners*** Listed below are ages of actresses and actors at the times that they won Oscars. The data are paired according to the years that they won. Use a 0.05 significance level to test the claim that there is no difference between the ages of best actresses and the ages of best actors at the time that the awards were presented.

Best Actresses	33	35	35	28	30	29	61	32	33	45
Best Actors	36	47	29	43	37	38	45	50	48	60

P–value: __________

Conclusion: ____

____

13-2. ***Oscar Winners*** Repeat the preceding exercise using the 82 pairs of ages listed in Data Set 11 in Appendix B of the textbook. Instead of manually entering the data, open the Minitab workbook OSCAR from the CD included with the textbook.

P–value: __________

Conclusion: ____

____

13-3. ***Flight Data*** Use the following times for American Airlines Flight 19 from New York (JFK) to Los Angeles (LAX). Use a 0.05 significance level to test the claim that there is no difference between taxi-out times and taxi-in times.

Taxi-Out Time (min)	15	12	19	18	21	20	13	15	43	18	17	19
Taxi-In Time (min)	10	10	16	13	9	8	4	3	8	16	9	5

P–value: __________

Conclusion: ____

____

13-4. ***Flight Data*** The preceding example uses flight data from American Airlines Flight 19, and Example 2 in Section 13-2 of the textbook uses the data from Flight 21. Data Set 15 in Appendix B includes sample data for 48 flights from Flights 1, 3, 19, and 21. Download the data from Data Set 15. The Minitab workbook name is FLIGHTS. Use the paired taxi-out times and taxi-in times from those 48 flights. Use a 0.05 significance level to test the claim that there is no difference between taxi-out times and taxi-in times.

P–value: __________

Conclusion: ____

____

13-5. ***Gender Selection*** The Genetics and IVF Institute conducted a clinical trial of its methods for gender selection. As of this writing, 239 of 291 babies born to parents using the YSORT method were boys. Use a 0.01 significance level to test the claim that the YSORT method has no effect.

P–value: __________

Conclusion: ____

____

13-6. ***Earthquake Magnitudes*** Refer to Data Set 16 in Appendix B for the earthquake magnitudes. The Minitab worksheet is named QUAKE. Use a 0.01 significance level to test the claim that the median is equal to 1.00.

P–value: __________

Conclusion: ____

____

13-7. ***Testing for Median Weight of Quarters*** Refer to Data Set 21 in Appendix B for the weights (in g) of randomly selected quarters that were minted after 1964. (The Minitab worksheet name is COINS.)The quarters are supposed to have a median weight of 5.670 g. Use a 0.01 significance level to test the claim that the median is equal to 5.670 g. Do quarters appear to be minted according to specifications?

P–value: __________

Conclusion: ____

____

Experiments 13–8 through 13–13, involve the **Wilcoxon signed-ranks test**.

13–8. ***Sign Test vs. Wilcoxon Signed–Ranks Test*** Repeat Experiment 13-1 by using the Wilcoxon signed-ranks test for matched pairs. Enter the Minitab results below, and compare them to the sign test results obtained in Experiment 13-1. Specifically, how do the results reflect the fact that the Wilcoxon signed-ranks test uses more information?

P–value: __________

Conclusion: ____

____

Comparison: ____

____

13–9. ***Sign Test vs. Wilcoxon Signed–Ranks Test*** Repeat Experiment 13-2 by using the Wilcoxon signed-ranks test for matched pairs. Enter the Minitab results below, and compare them to the sign test results obtained in Experiment 13-2. Specifically, how do the results reflect the fact that the Wilcoxon signed-ranks test uses more information?

P–value: __________

Conclusion: ____

____

Comparison: ____

____

13–10. ***Sign Test vs. Wilcoxon Signed–Ranks Test*** Repeat Experiment 13-3 by using the Wilcoxon signed-ranks test for matched pairs. Enter the Minitab results below, and compare them to the sign test results obtained in Experiment 13-3. Specifically, how do the results reflect the fact that the Wilcoxon signed-ranks test uses more information?

P–value: __________

Conclusion: ____

Comparison: ____

____

13–11. ***Sign Test vs. Wilcoxon Signed–Ranks Test*** Repeat Experiment 13-4 by using the Wilcoxon signed-ranks test for matched pairs. Enter the Minitab results below, and compare them to the sign test results obtained in Experiment 13-4. Specifically, how do the results reflect the fact that the Wilcoxon signed-ranks test uses more information?

P–value: __________

Conclusion: ____

____

Comparison: ____

____

13-12. ***Earthquakes*** Refer to Data Set 16 in Appendix B for the sample of paired earthquake magnitudes and depths. The Minitab worksheet is named QUAKE. Use a 0.10 significance level with the Wilcoxon signed-ranks test to test the claim of no difference between the magnitudes and the depths.

P–value: __________

Conclusion: ____

____

What is fundamentally wrong with this analysis?

____

13-13. ***Coke Contents*** Refer to Data Set 19 in Appendix B for the amounts (in oz) in cans of regular Coke. The Minitab worksheet is named COLA. The cans are labeled to indicate that the contents are 12 oz of Coke. Use a 0.05 significance level with the Wilcoxon signed-ranks test to test the claim that cans of Coke are filled so that the median amount is 12 oz.

P–value: __________

Conclusion: ____

____

In Experiments 13–14 through 13–17, use the **Wilcoxon rank-sum test**.
(Minitab's Mann-Whitney test procedure is equivalent to the Wilcoxon rank-sum test.)

13-14. ***IQ and Lead Exposure*** Data Set 5 in Appendix B lists full IQ scores for a random sample of subjects with medium lead levels in their blood and another random sample of subjects with high lead levels in their blood. (The Minitab worksheet is named IQLEAD.) Use a 0.05 significance level with the Wilcoxon rank-sum test to test the claim that subjects with medium lead levels have full IQ scores with a higher median than the median full IQ score for subjects with high lead levels.

P–value: __________

Conclusion: ____

13-15. ***Weights of Coke*** Data Set 19 in Appendix B lists weights (lb) of the cola in cans of regular Coke and diet Coke. (The Minitab worksheet name is COLA.) Use a 0.05 significance level with the Wilcoxon rank-sum test to test the claim that the samples are from populations with the same median.

P–value: __________

Conclusion: ____

____

13-16. ***Cigarettes*** Refer to Data Set 10 in Appendix B for the amounts of nicotine (in mg per cigarette) in the sample of king size cigarettes, which are non-filtered, non-menthol, and non-light, and for the amounts of nicotine in the 100 mm cigarettes, which are filtered, non-menthol, and non-light. (The Minitab worksheet name is CIGARET.) Use a 0.01 significance level with the Wilcoxon rank-sum test to test the claim that the median amount of nicotine in the non-filtered king size cigarettes is greater than the median amount of nicotine in the 100 mm filtered cigarettes.

P–value: __________

Conclusion: ____

____

13-17. ***Cigarettes*** Refer to Data Set 10 in Appendix B for the amounts of tar in the sample of king size cigarettes, which are non-filtered, non-menthol, and non-light, and for to the amounts of tar in the 100 mm cigarettes, which are filtered, non-menthol, and non-light. (The Minitab worksheet name is CIGARET.) Use a 0.01 significance level with the Wilcoxon rank-sum test to test the claim that the median amount of tar in the non-filtered king size cigarettes is greater than the median amount of tar in the 100 mm filtered cigarettes.

P–value: __________

Conclusion: ____

____

In Experiments 13–18 through 13–22, use the **Kruskal-Wallis test**.

13–18. ***Do All Colors of M&Ms Weigh the Same?*** Refer to the M&M weights in Data Set 20 in Appendix B from the textbook. (The Minitab worksheet name is M&M.) Use a 0.05 significance level with the Kruskal-Wallis test to test the claim that the weights of M&Ms have the same median for each of the six different color populations.

P–value: __________

Conclusion: ____

____

13-19. ***Nicotine in Cigarettes*** Refer to Data Set 10 in Appendix B and use the amounts of nicotine (mg per cigarette) in the king size cigarettes, the 100 mm menthol cigarettes, and the 100 mm non-menthol cigarettes. (The Minitab worksheet name is CIGARET.) The king size cigarettes are non-filtered, non-menthol, and non-light. The 100 mm menthol cigarettes are filtered and non-light. The 100 mm non-menthol cigarettes are filtered and non-light. Use a 0.05 significance level with the Kruskal-Wallis test to test the claim that the three categories of cigarettes yield the same median amount of nicotine.

P–value: __________

Conclusion: ____

____

13-20. ***Tar in Cigarettes*** Refer to Data Set 10 in Appendix B and use the amounts of tar (mg per cigarette) in the three categories of cigarettes described in the preceding exercise. Use a 0.05 significance level with the Kruskal-Wallis test to test the claim that the three categories of cigarettes yield the same median amount of tar.

P–value: __________

Conclusion: ____

____

13-21. ***Carbon Monoxide from Cigarettes*** Refer to Data Set 10 in Appendix B and use the amounts of carbon monoxide (mg per cigarette) in the three categories of cigarettes described in Exercise 19. Use a 0.05 significance level with the Kruskal-Wallis test to test the claim that the three categories of cigarettes yield the same median amount of carbon monoxide.

P–value: __________

Conclusion: ____

____

13-22. ***Passive and Active Smoke*** Data Set 9 in Appendix B of the textbook lists measured cotinine levels from a sample of subjects who smoke, another sample of subjects who do not smoke but are exposed to environmental tobacco smoke, and a third sample of subjects who do not smoke and are not exposed to environmental tobacco smoke. (The Minitab worksheet name is Cotinine.) Cotinine is produced when the body absorbs nicotine. Use a 0.01 significance level to test the claim that the three samples are from populations with the same median.

P–value: __________

Conclusion: ____

____

Experiments 13–23 through 13–28 involve **rank correlation.**

13-23. ***Judges of Marching Bands*** Two judges ranked seven bands in the Texas state finals competition of marching bands (Coppell, Keller, Grapevine, Dickinson, Poteet, Fossil Ridge, Heritage), and their rankings are listed below (based on data from the University Interscholastic League). Test for a correlation between the two judges. Do the judges appear to rank about the same or are they very different?

Band	Cpl	Klr	Grp	Dck	Ptt	FR	Her
First Judge	1	3	4	7	5	6	2
Second Judge	6	4	5	1	3	2	7

Rank Correlation Coefficient: __________

P–value: __________

Conclusion: ____

____

13-24. ***Judges of Marching Bands*** In the same competition described in the preceding experiment, a third judge ranked the bands with the results shown below. Test for a correlation between the first and third judges. Do the judges appear to rank about the same or are they very different?

Band	Cpl	Klr	Grp	Dck	Ptt	FR	Her
First Judge	1	3	4	7	5	6	2
Third Judge	3	4	1	5	7	6	2

Rank Correlation Coefficient: __________

P–value: __________

Conclusion: ____

____

13-25. ***Blood Pressure*** Refer to the measured systolic and diastolic blood pressure measurements of 40 randomly selected males in Data Set 1 in Appendix B of the textbook and test the claim that among men, there is a correlation between systolic blood pressure and diastolic blood pressure. (The Minitab worksheet name is MBODY.)

Rank Correlation Coefficient: __________

P–value: __________

Conclusion: ____

____

13-26. ***Blood Pressure*** Refer to the measured systolic and diastolic blood pressure measurements of 40 randomly selected females in Data Set 1 in Appendix B of the textbook and test the claim that among men, there is a correlation between systolic blood pressure and diastolic blood pressure. (The Minitab worksheet name is FBODY.)

Rank Correlation Coefficient: __________

P–value: __________

Conclusion: ____

____

13-27. ***IQ and Brain Volume*** Refer to Data Set 6 in Appendix B of the textbook and test the claim that there is a correlation between brain volume and IQ socre. (The Minitab worksheet name is IQBRAIN.)

Rank Correlation Coefficient: __________

P–value: __________

Conclusion: ____

____

13-28. ***Earthquakes*** Refer to Data Set 16 in Appendix B of the textbook and test the claim that there is a correlation between magnitudes and depths of earthquakes. (The Minitab worksheet name is QUAKE.)

Rank Correlation Coefficient: __________

P–value: __________

Conclusion: ____

____

Experiments 13–29 through 13–33 involve the **runs test for randomness**.

13-29. ***Oscar Winners*** Listed below are the genders of the younger winner in the Academy Awards categories of Best Actor and Best Actress for recent and consecutive years. Do the genders of the younger winners appear to occur randomly?

F F F M M F F F F F F M F F F M F F F

P–value: __________

Conclusion: ____

13-30. ***Testing for Randomness of Presidential Election Winners*** The political parties of the winning candidates for a recent sequence of presidential elections are listed below, where D denotes Democratic party and R denotes Republican party. Does it appear that we elect Democrat and Republican candidates in a random sequence?

R R D R D R R R R D D R R R D D D
D D R R D D R R D R R R D D R R D

P–value: __________

Conclusion: ____

13-31. ***Baseball World Series Victories*** Test the claim that the sequence of World Series wins by American League and National League teams is random. Given below are recent results, with American League and National League teams represented by A and N, respectively.

A N A N N N A A A A N A A A A N A N N A A N N A A A A N A N

N A A A A A N A N A N A N A A A A A A A N N A N A N N A A N

N N A N A N A N A A A N N A A N N N N A A A N A N A N A A A

N A N A A A N A N A A N A N A N

P–value: __________

Conclusion: ____

13-32. ***Stock Market*** Listed below are the annual high values of the Dow Jones Industrial Average for a recent sequence of years (as of this writing). Test for randomness below and above the median.

969	995	943	985	969	842	951	1036	1052	892
882	1015	1000	908	898	1000	1024	1071	1287	1287
1553	1956	2722	2184	2791	3000	3169	3413	3794	3978
5216	6561	8259	9374	11568	11401	11350	10635	10454	10855
10941	12464	14198	13279	10580	11625				

P–value: __________

Conclusion: ____

14

Statistical Process Control

14-1 Run Charts

In the Statistical Process Control chapter (Chapter 14) in the Triola textbook (excluding *Essentials of Statistics*) we define **process data** to be data arranged according to some time sequence, such as the data in the table below. The Chapter Problem for Chapter 14 in the textbook states that the measurements are from a new minting process. On each of 20 consecutive days of production, a quarter is selected during each of the first five hours of production and the quarter is weighed. The results are listed in the table.

Weights (grams) of Minted Quarters

Day	Hour 1	Hour 2	Hour 3	Hour 4	Hour 5	$\bar{x}$	s	Range
1	5.543	5.698	5.605	5.653	5.668	5.6334	0.0607	0.155
2	5.585	5.692	5.771	5.718	5.720	5.6972	0.0689	0.186
3	5.752	5.636	5.660	5.680	5.565	5.6586	0.0679	0.187
4	5.697	5.613	5.575	5.615	5.646	5.6292	0.0455	0.122
5	5.630	5.770	5.713	5.649	5.650	5.6824	0.0581	0.140
6	5.807	5.647	5.756	5.677	5.761	5.7296	0.0657	0.160
7	5.686	5.691	5.715	5.748	5.688	5.7056	0.0264	0.062
8	5.681	5.699	5.767	5.736	5.752	5.7270	0.0361	0.086
9	5.552	5.659	5.770	5.594	5.607	5.6364	0.0839	0.218
10	5.818	5.655	5.660	5.662	5.700	5.6990	0.0689	0.163
11	5.693	5.692	5.625	5.750	5.757	5.7034	0.0535	0.132
12	5.637	5.628	5.646	5.667	5.603	5.6362	0.0235	0.064
13	5.634	5.778	5.638	5.689	5.702	5.6882	0.0586	0.144
14	5.664	5.655	5.727	5.637	5.667	5.6700	0.0339	0.090
15	5.664	5.695	5.677	5.689	5.757	5.6964	0.0359	0.093
16	5.707	5.890	5.598	5.724	5.635	5.7108	0.1127	0.292
17	5.697	5.593	5.780	5.745	5.470	5.6570	0.1260	0.310
18	6.002	5.898	5.669	5.957	5.583	5.8218	0.1850	0.419
19	6.017	5.613	5.596	5.534	5.795	5.7110	0.1968	0.483
20	5.671	6.223	5.621	5.783	5.787	5.8170	0.2380	0.602

A *run chart* is a sequential plot of *individual* data values over time. A run chart can be generated as follows.

Minitab Procedure for Generating a Run Chart

1. Enter all of the data (in sequence) in column C1.
2. Select the main menu item of **Stat**.
3. Select the subdirectory item of **Quality Tools**.
4. Select the option of **Run Chart**.
5. Make these entries in the dialog box:
   - Select **Single column** and enter C1 in the adjacent box.
   - Enter 1 for the subgroup size (because we want *individual* values plotted).
   - Click **OK**.

Using the process data in the preceding table, the above procedure will result in this run chart:

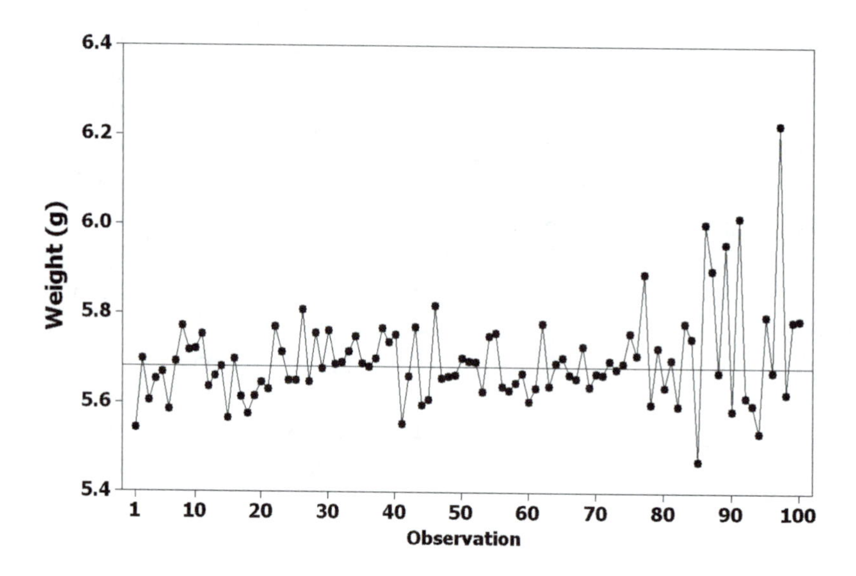

Examine the above run chart and note that it reveals this problem: As time progresses from left to right, the heights of the points appear to show a pattern of increasing variation. See how the points at the left fluctuate considerably less than the points farther to the right. It appears that the manufacturing process started out well, but deteriorated as time passed. If left alone, this minting process will continue to deteriorate because the run chart suggests that the process is not **statistically stable** (or **within statistical control**) because it has a pattern of increasing variation.

14-2 *R* Charts

The textbook describes *R* charts as sequential plots of ranges. Using the data from the preceding table, for example, the *R* chart is a plot of the ranges 0.155, 0.186, 0.187, . . . , 0.602. *R* charts are used to monitor the *variation* of a process. The procedure for obtaining an *R* chart is as follows.

Minitab Procedure for Creating *R* Charts

1. Enter all of the data in sequential order in column C1.

2. Select **Stat** from the main menu.

3. Select **Variable Charts for** Subgroups.

4. Select the option of **R**.

5. Make these entries in the dialog box:

    - Enter C1 in the box for the column containing all observations.

    - Enter a subgroup size of 5 in the indicated box. (We use 5 here because the preceding table of weights includes 5 observations each day.)

    - Click the **R options** button, click on the Estimate tab, and click on the **Rbar** button. Click on **Estimate**, then click on the small circle labeled **Rbar**. Click **OK**.)

Using the data in the preceding table, the Minitab display will be as shown below.

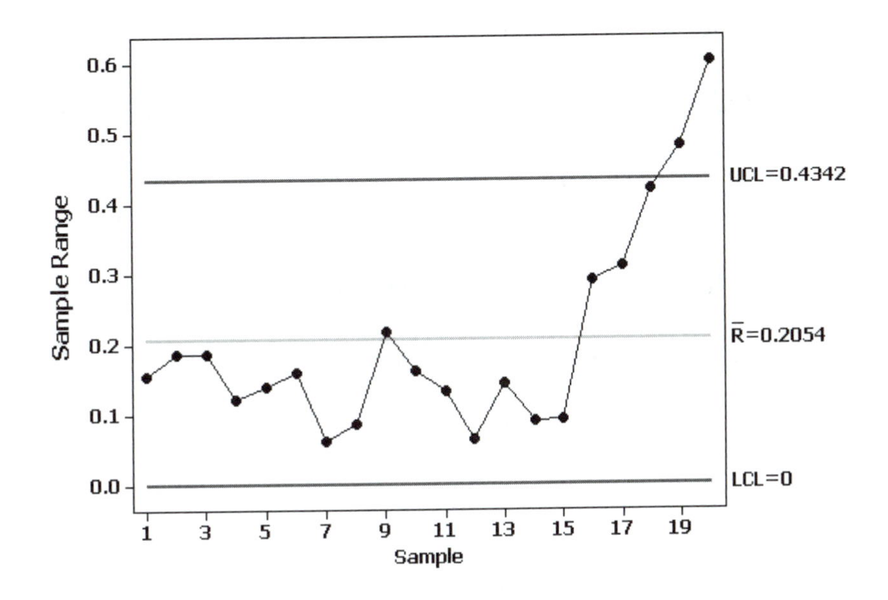

We can interpret the above *R* chart by applying these three criteria for determining whether a process is within statistical control:

1. There is no pattern, trend, or cycle that is obviously not random.
2. No point lies beyond the upper or lower control limits.
3. There are not 8 consecutive points all above or all below the center line.

Analyzing the above *R* chart leads to the conclusion that the process *is out of statistical control* because the second criterion is violated: There is a point beyond the upper control limit. Also, the third criterion is violated because there is a pattern of an upward trend, indicating that the process is experiencing increasing variation.

14-3 $\overline{x}$ Charts

We can now proceed to use the same weights of quarters to create an $\overline{x}$ chart. The textbook explains that an $\overline{x}$ chart is used to monitor the *mean* of the process. It is obtained by plotting the sample means. Follow these steps to generate an $\overline{x}$ chart.

Minitab Procedure for Generating an $\overline{x}$ Chart

1. Enter all of the data in sequential order in column C1.
2. Select **Stat** from the main menu.
3. Select **Variable Charts for** Subgroups.
4. Select the option of **Xbar**.
5. Make these entries in the dialog box:
    - Enter C1 in the box for the column containing all observations.
    - Enter a subgroup size of 4 in the indicated box. (We use 4 here because the data in the table of altimeter errors includes 4 observations each day.)
    - Click **Xbar options**, click the **Estimate** tab, and click on the **Rbar** button.

Using the data in the table of weights of minted quarters given earlier, the Minitab display will be as shown below.

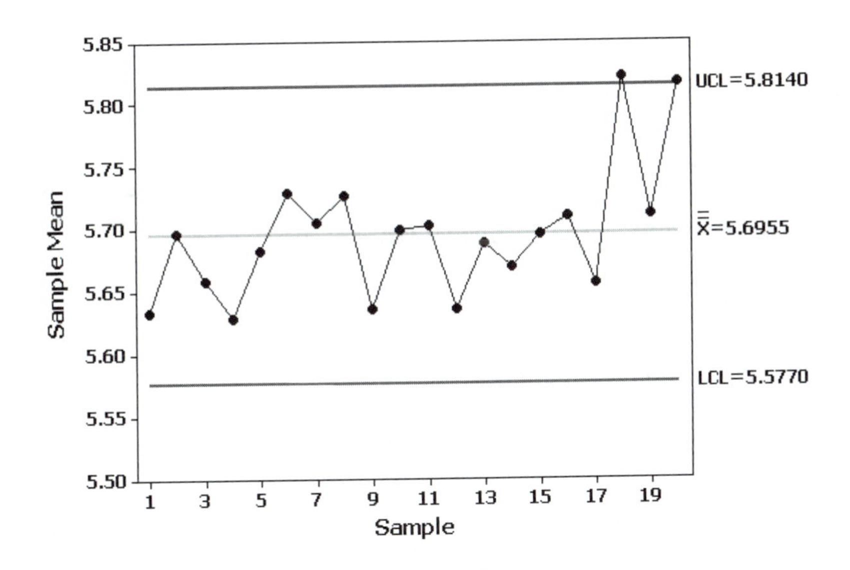

We can interpret the $\overline{x}$ chart by applying the three out-of-control criteria given in the textbook. We conclude that the mean in this process is *out of statistical control* because there is at least one point lying beyond the upper control limit.

14-4 *p* Charts

A control chart for attributes (or *p* chart) can also be constructed by using the same procedure for *R* charts and $\overline{x}$ charts. A *p* chart is very useful in monitoring some process proportion, such as the proportions of defects over time. See Example 1 in Section 14-3 in the textbook. Listed below are the numbers of defects in batches of 10,000 randomly selected each day from a new manufacturing process that is being tested.

Defects: 8 7 12 9 6 10 10 5 15 14 12 14 9 6 16 18 20 19 18 24

Follow these steps to generate a Minitab *p* chart:

Minitab Procedure for Generating a *p* Chart

1. Enter the numbers of defects (or items with any particular attribute) in column C1.
2. Select the main menu item of **Stat**.
3. Select the subdirectory item of **Control Charts**.
4. Select the option of **Attribute** Charts.
5. Make these entries in the dialog box:
   - Enter C1 in the box labeled Variable.
   - Enter the subgroup size. (For example, if C1 consists of the numbers of defects per batch and each batch consists of 10,000 quarters, then enter 10000 for the subgroup size.)
   - Click **OK**.

The Minitab display will be as shown below. We can interpret the control chart for p by considering the three out-of-control criteria listed in the textbook. Using those criteria, we conclude that this process is out of statistical control for this reason: There appears to be an upward trend; also, there is a point lying beyond the upper control limit. Immediate action should be taken to correct the increasing proportion of defects.

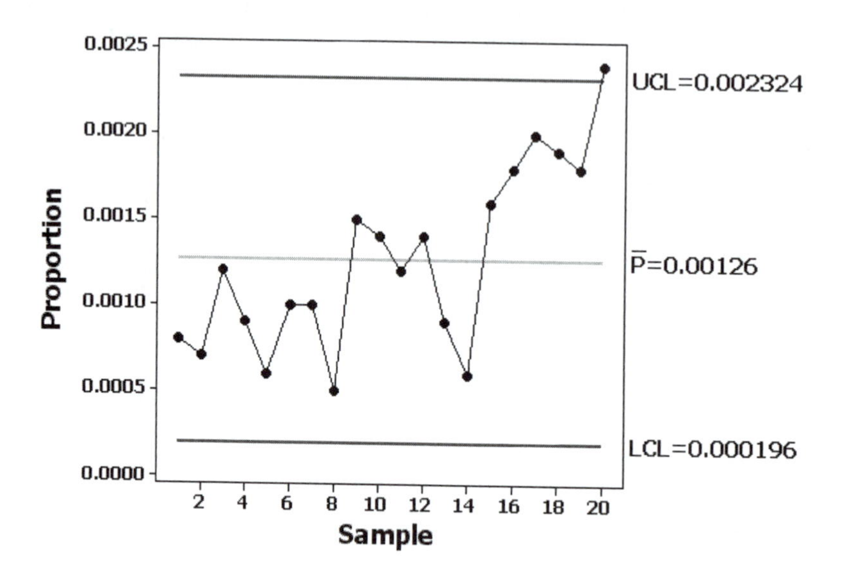

CHAPTER 14 EXPERIMENTS: Statistical Process Control

Constructing Control Charts for Aluminum Cans *Experiments 1 and 2 are based on the axial loads(in pounds) of aluminum cans that are 0.0109 in. thick, as listed in the "Axial Loads of Aluminum Cans" data set in Appendix B of the textbook. An axial load of a can is the maximum weight supported by its side, and it is important to have an axial load high enough so that the can isn't crushed when the top lid is pressed into place. The data are from a real manufacturing process, and they were provided by a student who used an earlier version of this book.*

14–1. ***R Chart*** On each day of production, seven aluminum cans with thickness 0.0109 in. were randomly selected and the axial loads were measured. The ranges for the different days are listed below, but they can also be found from the values given in Appendix B of the textbook. (The Minitab worksheet is *CANS.*) Construct an *R* chart and determine whether the process variation is within statistical control. If it is not, identify which of the three out-of-control criteria lead to rejection of statistically stable variation.

78 77 31 50 33 38 84 21 38 77 26 78 78
17 83 66 72 79 61 74 64 51 26 41 31

14–2. $\bar{x}$ ***Chart*** On each day of production, seven aluminum cans with thickness 0.0109 in. were randomly selected and the axial loads were measured. The means for the different days are listed below, but they can also be found from the values given in Appendix B of the textbook. (The Minitab worksheet is *CANS.*) Construct an $\bar{x}$ chart and determine whether the process mean is within statistical control. If it is not, identify which of the three out-of-control criteria lead to rejection of statistically stable variation.

252.7 247.9 270.3 267.0 281.6 269.9 257.7 272.9 273.7 259.1
275.6 262.4 256.0 277.6 264.3 260.1 254.7 278.1 259.7 269.4
266.6 270.9 281.0 271.4 277.3

Energy Consumption. *In Exercises 14-3 through 14-5, refer to Data Set 18 in Appendix B and use the measured voltage amounts for the power supplied directly to the author's home. (The Minitab worksheet name is VOLTAGE.) Let each subgroup consist of the five amounts within the business days of a week, so the first five voltages constitute the first subgroup, the second five voltages constitute the second subgroup, and so on. The result is eight subgroups with five values each.*

14-3. ***Home Voltage:*** $\bar{x}$ ***Chart*** Using subgroups of five voltage amounts, construct an $\bar{x}$ chart and determine whether the process mean is within statistical control. If it is not, identify which of the three out-of-control criteria lead to rejection of a statistically stable mean.

14-4. ***Home Voltage: Run Chart*** Construct a run chart for the 40 voltage amounts. Does there appear to be a pattern suggesting that the process is not within statistical control?

14-5. ***Home Voltage: R Chart*** Using subgroups of five voltage amounts, construct an R chart and determine whether the process variation is within statistical control. If it is not, identify which of the three out-of-control criteria lead to rejection of statistically stable variation.

Energy Consumption. *In Experiments 14-6 through 14-8, use the following amounts of electricity consumed (in kWh) in the author's home. Let each subgroup consist of the six amounts within the same year, so that there are eight subgroups with six amounts in each subgroup.*

Year 1	3637	2888	2359	3704	3432	2446
Year 2	4463	2482	2762	2288	2423	2483
Year 3	3375	2661	2073	2579	2858	2296
Year 4	2812	2433	2266	3128	3286	2749
Year 5	3427	578	3792	3348	2937	2774
Year 6	3016	2458	2395	3249	3003	2118
Year 7	4261	1946	2063	4081	1919	2360
Year 8	2853	2174	2370	3480	2710	2327

14-6. ***Energy Consumption: R Chart*** Let each subgroup consist of the 6 values within a year. Construct an R chart and determine whether the process variation is within statistical control. If it is not, identify which of the three out-of-control criteria lead to rejection of statistically stable variation.

14-7. ***Energy Consumption: $\bar{x}$ Chart*** Let each subgroup consist of the 6 values within a year. Construct an $\bar{x}$ chart and determine whether the process mean is within statistical control. If it is not, identify which of the three out-of-control criteria lead to rejection of a statistically stable mean.

14-8. ***Energy Consumption: Run Chart*** Construct a run chart for the 48 values. Does there appear to be a pattern suggesting that the process is not within statistical control?

14-9. ***p Chart for Defective Defibrillators*** Consider a process that includes careful testing of each manufactured defibrillator. Listed below are the numbers of defective defibrillators in successive batches of 10,000. Construct a control chart for the proportion p of defective defibrillators and determine whether the process is within statistical control. If not, identify which of the three out-of-control criteria apply.

Defects: 20 14 22 27 12 12 18 23 25 19 24 28 21 25 17 19 17 22 15 20

14-10. ***p Chart for Defective Defibrillators*** Repeat the preceding exercise assuming that the size of each batch is 100 instead of 10,000. Compare the control chart to the one found for the preceding exercise. Comment on the general quality of the manufacturing process described in the preceding exercise compared to the manufacturing process described in this exercise.

14-11. ***Violent Crimes*** In each of recent and consecutive years, 100,000 people in the United States were randomly selected and the number who were victims of violent crime was determined, with the results listed below. Does the rate of violent crime appear to exhibit acceptable behavior? (The values are based on data from the U.S. Department of Justice, and they are the most recent values available at the time of this writing.) How does the result affect us?

685 637 611 566 523 507 505 494 476 463 469 474 458 429

14-12. ***Cola Cans*** In ach of several consecutive days of production of cola cans, 500 cans are tested and the numbers of defects each day are listed below. What action should be taken?

20 22 19 17 19 15 16 13 14 14 11 13 12 11 10 9 9 10 7 7

Minitab Release 16 Important Menu Items

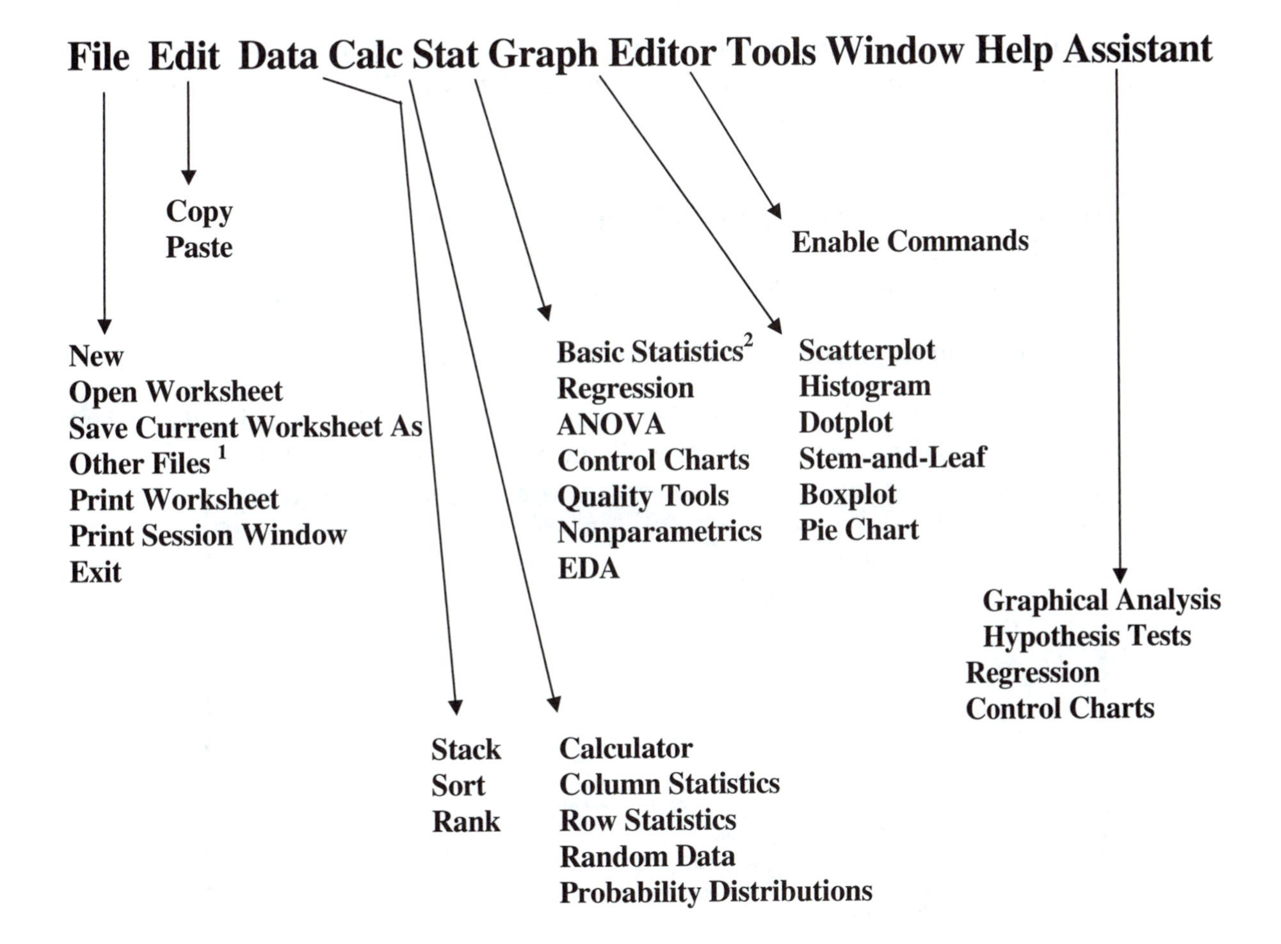

[1] Other Files: Use for importing and exporting text files.

[2] Basic Statistics: Descriptive Statistics
Confidence intervals and hypothesis tests for one or two proportions
Confidence intervals and hypothesis tests for one or two means
Confidence intervals and hypothesis tests for matched pairs
Hypothesis test for equality of two variances
Linear correlation
Test for normality

Downloading Worksheets from the CD

The CD-ROM included with the Triola textbook has worksheets for the data in Appendix B of the textbook. Appendix B in the textbook lists the worksheet names, and those names are also listed on the following page. Here is the procedure for retrieving a worksheet from the CD-ROM:

1. Click on Minitab's main menu item of **File**, then click on **Open Worksheet**.

2. You will get a dialog box like the one shown below. In the "Look in" box at the top, select the location of the stored worksheets. For example, if the CD-ROM is in drive D and you want to open the Minitab worksheet POTUS, do this:

   - In the "Look in" box, select drive **D** (or whatever drive contains the CD-ROM).
   - Double click on the folder **App B Data Sets**.
   - Double click on the folder **Minitab**.
   - Click on **POTUS** (or any other worksheet that you want), as shown below.

3. Click on the **Open** bar located in the lower right portion of the dialog box.

4. You will get a message that a copy of the contents of the file will be added to the current project. Click **OK**. The columns of data will now be in the Minitab worksheet.

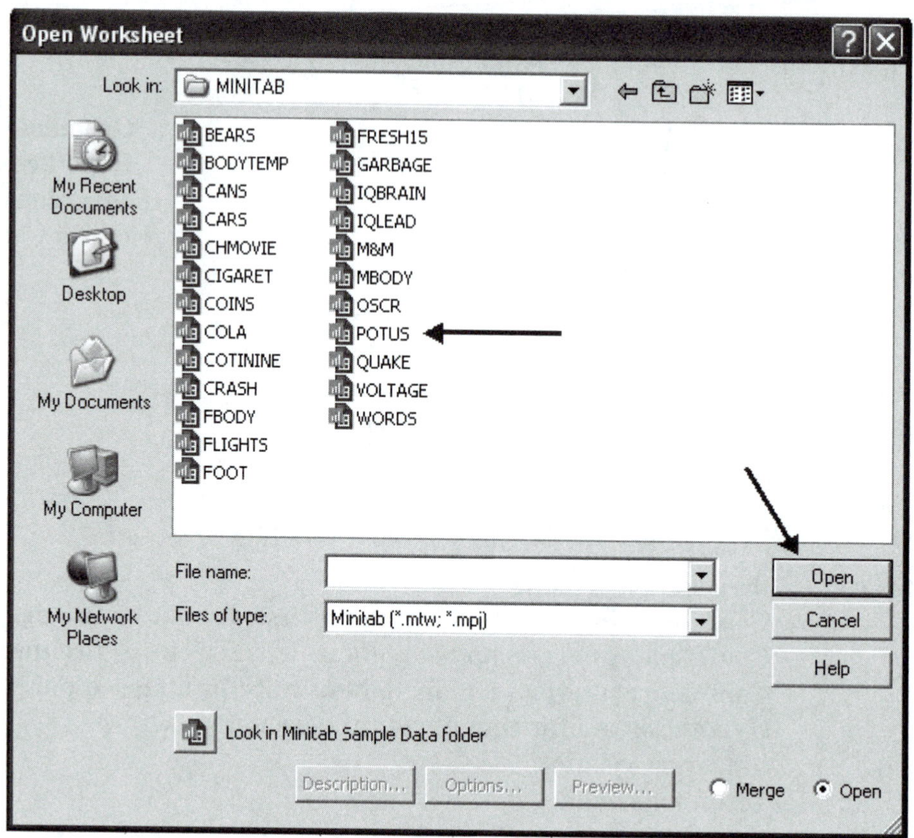

Appendix B: Data Sets

		Minitab Worksheet Name
Data Set 1:	Body Measurements	**MBODY and FBODY**
Data Set 2:	Foot and Height Measurements	**FOOT**
Data Set 3:	Body Temperatures of Healthy Adults	**BODYTEMP**
Data Set 4:	Freshman 15 Data	**FRESH15**
Data Set 5:	IQ and Lead Exposure	**IQLEAD**
Data Set 6:	IQ and Brain Size	**IQBRAIN**
Data Set 7:	Bear Measurements	**BEARS**
Data Set 8:	Alcohol and Tobacco Use in Animated Children's Movies	**CHMOVIE**
Data Set 9:	Passive and Active Smoke	**COTININE**
Data Set 10:	Cigarette Tar, Nicotine, and Carbon Monoxide	**CIGARET**
Data Set 11:	Ages of Oscar Winners	**OSCR**
Data Set 12	POTUS	**POTUS**
Data Set 13:	Car Crash Tests	**CRASH**
Data Set 14:	Car Measurements	**CARS**
Data Set 15:	Flight Data	**FLIGHTS**
Data Set 16:	Earthquakes	**QUAKE**
Data Set 17:	Word Counts by Males and Females	**WORDS**
Data Set 18:	Voltage Measurements from a Home	**VOLTAGE**
Data Set 19:	Weights and Volumes of Cola	**COLA**
Data Set 20:	M&M Plain Candy Weights	**M&M**
Data Set 21:	Coin Weights	**COINS**
Data Set 22:	Axial Loads of Aluminum Cans	**CANS**
Data Set 23:	Weights of Discarded Garbage	**GARBAGE**

Index